Jamie James was born and raised in Houston, Texas, and graduated from Williams College. Upon graduation he wrote a regular column for *Andy Warhol's Interview*. Since then, he has written about music, science, art, archaeology, and other subjects in the pages of many American and British magazines and newspapers. Mr. James is co-author, with David Soren, of *Kourion: The Search for a Lost Roman City*, a book about the archaeology of Cyprus, which was published in 1988. He is also the co-author, with archaeologists Russell Ciochon and John Olsen, of *Other Origins*, the account of a National Geographic-sponsored excavation in Vietnam. Mr. James lives in Bali, Indonesia.

The Music of the Spheres

JAMIE JAMES

The Music of the Spheres

Music, Science, and the Natural Order of the Universe

ABACUS

First published in the United States of America by Grove Press 1993
First published in Great Britain by Little, Brown and Company 1994
This paperback edition published by Abacus 1995

11 13 15 14 12 10

Copyright © Jamie James 1993

The moral right of the author has been asserted.

A CIP catalogue record for this book
is available from the British Library.

ISBN 978-0-349-10542-0

Printed and bound in Great Britain by
Clays Ltd, St Ives plc

Papers used by Abacus are from well-managed forests
and other responsible sources.

MIX
Paper from
responsible sources
FSC   FSC® C104740
www.fsc.org

Abacus
An imprint of
Little, Brown Book Group
Carmelite House
50 Victoria Embankment
London EC4Y 0DZ

An Hachette UK Company
www.hachette.co.uk

www.littlebrown.co.uk

For Archi

Acknowledgments

For their assistance in supplying illustrations for this book, I would like to thank the Bancroft Library, University of California, Berkeley; the Fellows of Magdalene College, Cambridge; the Rare Book Room of the New York Public Library; the Bibliothèque Nationale, Paris; Professor James Stevens Curl; and the Arnold Schoenberg Institute. Thanks also to Lionel Salter, for permission to quote from his translation of *Il sogno di Scipione*.

At an early stage in the planning of this book, Joscelyn Godwin suggested some important lines of research that proved to be very helpful. I am especially grateful to Janet Byrne, Christopher Hogwood, and Bob James, who read the early versions of the manuscript and offered some valuable advice. Martha Rowen and Maria Grazia Praderio helped me with some of the translations, but any errors there or elsewhere are my own. Heartfelt thanks to Mark Livingston for putting me up in Berkeley, and to Josef Astor for the photographs.

Contents

Contents

Illustrations

Preface

When pressed to describe this book in two lines of newspaper type while it was still being written, I have said that it is about "science and music." While that phrase has the virtue of both fitting on two lines and making sense, it has the disadvantage of not being a very accurate description of this book. It is not entirely misleading; you will encounter scores of musicians and scientists in these pages, sometimes behaving in ways that will surprise you. Yet the book itself has a much more limited, and I hope more reasonable, focus: I have attempted to cover the area of overlap between music and science, beginning at the beginning of Western civilization, when the two were identified so profoundly that anyone who suggested that there was any essential difference between them would have been considered an ignoramus, and arriving at the present day, when someone proposing that they have anything in common runs the risk of being labelled a philistine by one group and a dilettante by the other—and, most damning of all, a popularizer by both.

I have worked throughout my professional life as a journalist both of music and of science, which, obviously, has brought me into contact with a great many musicians and scientists. I never cease to be amazed and dismayed by how many of the former have no idea who Johannes Kepler was and how many of the latter have not read a novel since their freshman year. Ever since C. P. Snow coined the phrase "the Two Cultures" in 1959, there has existed a tiny club of people who regularly denounce this psychotic bifurcation in our civilization. Any number of scholarly panels and earnest speeches to matriculating undergraduates have been devoted to this deplorable state of affairs, yet it has not made the slightest difference. In this era of over-specialization, what passes for a Renaissance man is a biologist who goes to the opera twice a year, or a poet who uses the laws of planetary motion as a metaphor for love.

I belong to that club, though I do not have any hope that the situation will ever change. In fact, although I am not an old man, in my own life I have seen the Two Cultures drift perceptibly farther apart. Science has become so clever at improving television reception and breeding healthy cows that it seems no longer to have much time left over for accomplishing its original aims; and while the most fundamental art and literature of our culture have been consigned to the obscure margins of the curriculum, the most obscure and marginal art has come to hold center stage among the shrinking group of people who care about such things.

It was not always that way. There was a time when the universe was believed to cohere, when human life had a meaning and purpose. A person who devoted himself to a lifetime of study, instead of coming out at the end of it the author of a definitive treatise on the pismire, or a catalogue of the references to Norse sagas in *Finnegans Wake*, would actually have a shot at discovering the key to the universe. I am not a believer in the golden age; people's lives in the times I am about to describe were full of all kinds of unpleasantness. Nor do I believe that humankind will ever again be able to return to that kind of intellectual certainty. The key to the universe is no longer of use to anyone, because the exquisite edifice it once unlocked has crumbled into nothingness. Nonetheless, it does seem worth knowing that down through the vastest majority of history, our ancestors believed that the world made sense, that it was a place where they belonged. And because they were human even when they were wrong, we can belong there, too.

We are all Pythagoreans.
—Iannis Xenakis

The Music of the Spheres

ONE

The Great Theme

Picture to yourself, if you can, a universe in which everything makes sense. A serene order presides over the earth around you, and the heavens above revolve in sublime harmony. Everything you can see and hear and know is an aspect of the ultimate truth: the noble simplicity of a geometric theorem, the predictability of the movements of heavenly bodies, the harmonious beauty of a well-proportioned fugue—all are reflections of the essential perfection of the universe. And here on earth, too, no less than in the heavens and in the world of ideas, order prevails: every creature from the oyster to the emperor has its place, preordained and eternal. It is not simply a matter of faith: the best philosophical and scientific minds have proved that it is so.

This is no New Age fantasy but our own world as scientists, philosophers, and artists knew it until the advent of the Industrial Revolution and its companion in the arts, the Romantic movement. Those ideals are gone forever. After the revelations of modern scientific enquiry, educated people will never again be able to face the universe, now unimaginably complex, with anything like the serenity and certitude that existed for most of our history.

3

The concepts of the musical universe and the Great Chain of Being originate in the classical bedrock of our culture, flow through the Christian tradition, and remain firmly centered in the Renaissance and the Age of Reason. They are at the core of the culture. It was not until the nineteenth century that the perspective shifted decisively to the earthly, the tangible. Materialism and sensuality, qualities that had been deeply mistrusted throughout most of the Western tradition, emerged ascendant.

All art, including music, was a much more serious matter before the self-conscious aestheticism of the late nineteenth century took root. It is a recent notion that music is a divertissement to be enjoyed in comfortable surroundings at the end of the day, far removed from the hurly-burly of life's business; throughout most of the history of our culture, music was itself an essential part of life's business. Its ability to give pleasure was deemed to be not only the least of its attributes but even a perversion of its true purpose. Most serious thinkers before the nineteenth century considered the sensual delight of musical performances to have the same relationship to the ideal nature of music that sex had to love in Christianity: the former were transitory, without higher purpose, and ultimately debilitating to the soul; the latter were pure and enlightening, providing a connection between our earthly existence and eternal reality.

In the same way, science was deeply involved in the whole ethos of the culture. The conception of science as an enterprise carried out by a specially educated elite, working in laboratories filled with mysterious machines, would have been incomprehensible to an educated person before the close of the last century. The history of science is the continuing process of the widening gulf between the ideals and the practice of science. At the birth of Western science in Greece in the sixth century B.C., the two were identical. The asking of the questions was the intellectual breakthrough, and the answers were as poetic and expansive as the questions, for there existed no data with which they were expected to conform, aside from the perceived order and beauty of creation. "Doing things" was disdained as unworthy of science, whose true purpose was to elucidate the fundamental unities that explain the function and thus the meaning of the phenomenal world.

As scientific observation accumulated information, ostensibly to make the answers to the questions more precise, the universe revealed itself to be far more complicated than anyone had ever imagined. The assumption throughout centuries of science had been that there was a logic underlying the apparent chaos of creation, but that human perception was too clouded and fallacious to discern it. By the nineteenth century science had abandoned that position, and the search for the fundamental unities became more and more a theoretical goal. An abrupt conceptual turnabout had taken place: whereas Plato had taught that anything the eye could see was illusory, modern science teaches that the only things that do exist are those that we can see and touch (even if we "see" them with radio signals, or "touch" them with remote robotic devices).

Nowadays most scientists would accept the thesis that the cosmos has no underlying logic in the classical sense, but is rather a confluence of accidents, which are governed by laws. However, the laws themselves are irrational and do not arise from any fundamental orderliness. The concept of the universe as a random, meaningless place was expressed on the earthly level by the theory of evolution: the mutations that determine the course of life on earth, and indeed the very creation of humankind, were revealed to be largely fortuitous events.

Beginning at the close of the nineteenth century, science emerged as a powerful institution. Increasingly, as it came to be identified with technological thaumaturgy and the improvement of the quality of life, science ceased to be a purely intellectual enterprise: too much money was now at stake. The scientist, who had previously held a position in society equivalent to that of the poet, was now pressed into service by government and industry to work more marvels. These technological accomplishments were useful, but science became the victim of its own success, at least insofar as its traditional aims were concerned. The need to "produce" is as deadly to pure science as it is to poetry. As society came to demand ever more ingenious machines, the search for the transcendental harmonies atrophied.

The paradox (again, an evil one from the classical point of view) is that pure science occupied a more honorable position in society before it became so useful to it. Astronomers traditionally held prestigious places at court in antiquity and in the Renaissance, just as did poets (though it should be conceded that the former had to devote a great deal of time to drawing up horoscopes for their royal patrons, while the latter were obliged to compose laudatory odes addressed to them, suitable for recitation at public ceremonies). Yet these state duties did not inhibit scientists from undertaking experimental work solely to please themselves; indeed, they were able to do so exactly because they were to a large extent set apart from society.

Of course, tremendous difficulties were presented to Western scientists in the Christian era by church doctrine, of which the trial of Galileo is only the most famous example. Yet the religious variable in the social algebra of science has remained constant: the resistance to evolutionary theory, which continues to the present day, and the hurdles being put in the way of medical researchers who would use fetal tissue, to name two examples, are essentially the same as the religiously motivated prejudices against heterodox science five hundred years ago.

The rigid hierarchism of the pre-industrial world, while it may have been oppressive to most people in the society, actually tended to enhance the creative freedom of scientists. In every age of great creativity that was supported by the established authority—in the reigns of Augustus, of the Medici, of Elizabeth I, of Louis XIV—the sovereign was secure. If there was economic distress, it did not much affect the court. No one ever put a gun to Leonardo da Vinci's head and said, "Double the corn crop, or it's back to the day job with you." So long as science itself was not invested with real power in the society, it had a correspondingly greater freedom to pursue knowledge for its own sake. In the first place there was no significant motive of competition; and in the second place, there were usually no grave penalties for being wrong, which is an essential component of all scientific work.

Until the Industrial Revolution, scientists were essentially outsiders, supported by classically educated patrons who took an interest in their work, to some extent for its own sake. If a useful application happened to come out of the work, so much the better, but that was never the sine qua non of science. Yet there can be little doubt that that is what it is today. There are very few prestigious grants or government posts to be won by thinking about the ultimate aims of science.

This reshaping of science is not simply a matter of its growing complexity and power. It is a fundamental change in course. By comparison, while we now consider medicine to be a part of science, it was originally thought to be an art. Although medicine has undergone changes as profound as any field of human endeavor, today it is dedicated to precisely the same ends, to heal the sick and to protect life, as it was at the beginning of civilization. The Hippocratic oath is still administered to doctors at the beginning of their careers, just as it was more than two thousand years ago. There have been some changes, of course: the oath originally bound those who swore it not to help a pregnant woman get an abortion, a provision that is now usually omitted.

Another example, briefly. The practice of law from Hammurabi to the present has been wholly transformed time after time, yet its aim has always been the same, to dispense justice, although there has never been a consensus as to what justice is, and there never will be. Our own system has as a basic principle the presumption of innocence, which would have seemed ridiculous to Hammurabi; on the other hand, his code contained provisions we moderns find barbarous and cruel. Yet Hammurabi's *aim* was not to be cruel but rather to be just. In order to protect itself from criminals and troublemakers, he might have said, society is entitled to exact swift and exemplary punishment.

Science, on the other hand, has changed in its essential nature. Whereas modern doctors still swear the Hippocratic oath, and whereas much of our law today is based directly upon very old traditions of English common law (and ultimately, indeed, upon the Code of Hammurabi), classical science plays a very small part in the education of today's scientists. The visionaries and philosophers who created Western science are seen as noble but quaint, endearing, their "mistakes" ingenious in their way but nonetheless slightly risible. As far as most scientists are concerned, their fields are being invented daily. You are only as good as your latest published paper, which is more likely to attract notice if it disproves a paper published by someone else last year. Although credit is scrupulously given where it is due, the basic message to graduate students in science is that yesterday's misconceptions are about as useful as yesterday's toast.

That is not to suggest that what we now call pure science is not going on. Of course there are scientists who are studying, for example, the origin of life on the planet; yet they keep the matter firmly in the realm of biochemistry. Any discussion of what this means about man's place in the universe, and why we came to be here in the first place, is regarded as an extraneous diversion. There are astronomers who devote themselves to studying the extent and age of the universe—again, only so long as it remains a question of interpreting radio signals. No "real" scientist would ever embroil himself in questions like these: "Life on this planet came into existence shortly after it was formed some 4.6 billion years ago; then for 2 billion years, more than half the time there has been life on earth, the most complex form it took was algae. Why did higher life forms emerge? Is it just a quirk of biochemistry that has made me a person rather than a cell of an algal mat?" Or: "The cosmos is 18 billion years old—excellent. But what came before that? Nothingness? Can something (that is, everything) come from nothing?" Those, of course, are the questions that we really want answered. But even the purest of the pure scientists would smile at the naïveté of such questions, and tell us that they are imponderables, unanswerable matters better left to the philosophers and theologians.

Of course such questions are eminently ponderable—we have been pondering them ever since we began scratching on clay tablets, perhaps since we started walking upright. What our hypothetical scientist means is that these questions cannot be pondered within the parameters of science as it is now practiced, because there is no answer provable with the degree of certainty required by current methods. Far from being unanswerable, the question of where the universe came from, and how man came to have a place in it, has a great many answers. That is exactly the problem: science demands that there be only one answer, indisputably right, which it can propagate through the university curriculum. Likewise, it is permissible for a biologist to discuss the emergence of life on this planet so long as he disdains ruminating about *why* this interesting event took place. Thus in the modern world the phrase *pure science* has a double meaning. Not only is it outside the bounds of applied science, it is positioned just as firmly this side of the realm of the ultimate, the transcendent, the absolute.

None of the foregoing ought to be construed as an attack on contemporary science: like any intellectual activity, science must be judged by its own criteria. When science became an institutionalized power in society, it became concerned above all else with being right, for a reason that is well known to any logician: a conclusion based upon faulty premises will itself prove to be faulty. Thus for a student of biochemistry or mathematics or astrophysics, it is essential that all the basic premises be absolutely correct.

This imperative of infallibility caused the questions to be rephrased. If every scientific problem must have an irrefutable mathematical solution, and if every mathematical expression is conceived as having no meaning outside itself, then the grey shades of subjectivity required to examine the fundamental, underlying questions posed by the perceptible universe render the questions themselves irrelevant. That is not to say that the questions now addressed by science are lacking in interest: the biochemical background to the emergence of life, for example, fascinates; and putting a number on the age of the cosmos is unquestionably a worthy intellectual pursuit. But it is sophistry to pretend that, having posed answers to them, one has resolved the basic questions that prey most forcibly on the human mind. Moreover, our industrial society now places less importance on these intellectual quests than it does on the biochemistry of cheap fuel, or the cosmology required to put telecommunications satellites into orbit.

Science has drifted so far from its original aims that even to bother with the question of its relationship to music might appear to be an exercise in irrelevancy, like chronicling the connection between military history and confectionery. Yet every scholar of the history of science or of music can attest to the intimate connection between the two. In the classical view it was not really a connection but an identity.

I have asserted that this phenomenon is widely accepted among historians of science. What is perhaps less well understood is that an analogous process has been at work in the evolution of music, which has been called the purest of the arts. By a process profoundly like that at work in the evolution of science, music has become established as a profane and temporal institution. As a result, the ends of music, its rationale for existing, has become muddled. The history of music, if it is to be anything more than a long and tedious succession of clefs and quavers, is the story of how musical ideals and musical practice have drifted ever further apart. For as music began to serve a variety of social functions—in the beginning to excite bravery in soldiers, for example, or to induce a state of mind receptive to religious experience—and ultimately as musical institutions themselves became rich and powerful fixtures in society, the practice of music, like that of science, gradually came to be alienated from the deeper questions posed by the very fact of its existence.

In the classical era, this underlying psychosis was exemplified by Plato's hostility toward μουσική (*mousikē*) (which, it ought to be borne in mind, meant any human activity governed by the Muses). While the philosopher would allow music that served the state (albeit reluctantly), whenever its sole purpose was to induce pleasure, he proscribed it. In the Middle Ages, this duality is expressed in the distinction, to be found in virtually every surviving scholarly treatise of the period, between the *musici*, music theorists who were considered to be the field's true artists, and *cantores*, mere musical performers, the singers and instrumentalists who were usually regarded as contemptible, immoral people. Over the course of time spanned by the music to which most of us listen, that is to say over the past four hundred years, this dualistic debate about the aims of music has been clouded and rephrased in irrelevant ways: sacred versus profane, high art versus folk art, traditional versus progressive, upper-case Classical versus Romantic, lower-case classical versus pop. Yet at bottom, the question has remained whether music has the power to "improve," to guide man toward an awareness and contemplation of a higher beauty, of an ultimate reality, or whether it exists only to beguile the hours pleasurably.

One result of these parallel processes in the history of music and science, which we may roughly characterize as the estrangement of their practice from their theoretical aims, has been to create an artificial divorcement between the two fields. In the modern age it is a basic assumption that music appeals directly to the soul (which has been called by many names—sensibility, temperament, the emotions, among others) and bypasses the brain altogether, while science operates in just the reverse fashion, confining itself to the realm of pure ratiocination and having no contact at all with the soul. Another way of stating this duality is to marshal on the side of music Oscar Wilde's dictum that "All art is quite useless," while postulating that science is the apotheosis of earthly usefulness, having no connection with anything that is not tangibly of this world.

These suppositions would have seemed very strange to an Athenian of Plato's day, to a medieval scholar, to an educated person of the Renaissance, even to a habitué of London's coffeehouses in the eighteenth century. They would all have taken it for granted that music served a number of quite specific purposes: to instil patriotism and piety, to give succor in times of distress, to cure the sickened soul, and so forth. Conversely, science was viewed as having a deeply spiritual element. By illuminating creation and the laws that governed it, by elucidating the order that underlay apparent chaos, the scientist was bringing into the world something quite as beautiful as the work of any painter, sculptor, or composer.

Then something happened. Two hundred years ago the world was turned upside down. The social order abruptly tumbled, a phenomenon symbolized by the falling stones of the Bastille, and simultaneously a revolution took place in what came to be called the humanities. The pious moralism of writers like Samuel Richardson and Samuel Johnson, the dominant literary personalities of eighteenth-century London, was supplanted by the idiosyncratic, even bizarre visions of Coleridge and Blake. In painting, the serene, sheltering arcadias of Watteau and Gainsborough were displaced by Goya's horrific allegories, Fuseli's nightmares, Delacroix's patriotic gore.

And in music there was Beethoven. Between the finale of Mozart's *Così fan tutte*, which in 1790 proclaimed, "Happy is the man who makes reason his guide," and the willful, majestic melancholy of Beethoven's late string quartets, thirty-five years later, we may trace a profound alteration in temperament. Gloated the German critic Ludwig Börne in 1830, writing about the young generation so profoundly influenced by Beethoven, "It is a joy to see how the industrious Romantics light a match to everything and tear it down, and push great wheelbarrowfuls of rules and Classical rubbish away from the scene of the conflagration." Of course the rules and rubbish he refers to are precisely the concepts of the orderly cosmos and the Great Chain of Being that had served as the cornerstones of Western thought since its inception.

Thus how supremely ironic it is that this revolutionary school of music, Romanticism, should today be enshrined as the official culture, almost to the exclusion of what preceded it and what was to come after. Works that literally incited revolution—in Wagner's Dresden, in Verdi's Milan—are now considered to be the stately theme music of established authority. Pieces that modern audiences listen to in hushed reverence sometimes provoked outrage at their premieres. Christoph von Dohnányi recently made this comment: "Everyone thinks that he understands Beethoven better than contemporary music. In fact, it is much more difficult to understand Beethoven than any modern composer you can name, without a great deal of study." We may interpret that to mean that unless we understand the musical world order that Beethoven was overturning, we shall never understand the meaning of his achievement. Yes, it is great music—but why? Not because it has lovely melodies, nor because its stirring rhythms stimulate nervous excitement. It is great music because it pointed the way to a new direction.

Yet for all its imagination and sheer musical power, Romanticism is an anomaly in the history of music. This music we were all taught to regard as the great music—the symphonies of the

nineteenth-century Middle European composers, Italian and
German opera of the same period—actually constitutes a relatively small sliver of the whole Western musical repertory.
Characterized by a high pitch of emotional expression and a
frequently anthropocentric "subject matter," the music of the
Romantics might have seemed like much pompous ado about
nothing to the composers, musicians, and audiences of the centuries, the millennia, that preceded it.

Yet the victory of the Romantics was absolute. The music
of their great standard-bearers ultimately triumphed in the
concert halls and opera houses of Europe and throughout the
New World, sweeping all before it (though that in itself is not
so extraordinary—after all, in the eighteenth century Handel
and Haydn, to name just two, had achieved astounding popular success in their lifetimes). In the Romantic age for the first
time music became a commercial institution, another important similarity with what was happening in the sciences. Composers before the rise of Romanticism, like court scientists,
only had to please one person, their royal or ecclesiastical
patron. However, beginning in Central Europe in the late
eighteenth century, concerts became public affairs, and for the
first time musicians had to deal with the phenomenon of the
box office. No longer was it enough simply to be good; one
had to be successful. Music publishing became a major force
in popularizing compositions for home consumption. In the
nineteenth century, many more people knew the operas of
Wagner and Verdi, for example, from piano redactions of their
scores than from experiences in the opera house. Later on, at
the beginning of the twentieth century, radio and the phonograph made Romantic music—now known as "classical
music"—a vast business institution. Just as was the case with
the rise of science as a social institution, once a great deal of
money was at stake, the criteria shifted irrevocably. If being
correct became the overriding imperative of science, the same
thing held true in concert halls—listening to the "right" music

has been a social obligation for the elite audience of classical music for the past hundred years.

One of the most remarkable corollaries of the triumph of Romanticism has been its powerful transforming effect on the compositions of preceding eras. In the nineteenth century Handel's oratorios and Bach's masses were shamelessly tarted up for elephantine public performances, lavishly re-scored and performed by forces far exceeding what their composers intended. This practice lasted well into the twentieth century, when even so powerful and highly respected a musical figure as Sir Thomas Beecham thought nothing of rewriting Baroque music to make it sound magnificently (and inappropriately) Romantic.

More subtle but not less pernicious is the way in which the Romantic sensibility has affected the way we hear the music of any era, even when the performance itself may be blameless in intention. Take as an example Henry Purcell's *Dido and Aeneas*, the earliest surviving English opera. While the opera is alive with passion, it is an exquisitely refined Mannerist passion, firmly grounded in the classical tradition, with a sophisticated irony that is almost sure to be lost on a modern audience. Dido's Lament, the opera's finale, is sublimely pathetic, but it is not sentimental. When the dying Carthaginian queen sings "Remember me, but ah! forget my fate," the audience is moved, but there is a dignity and nobility to the emotional response that is worlds away from the exaggerated self-pity of, say, Puccini's Manon Lescaut when she sings "Sola, perduta, abbandonata" in an equally pathetic situation. A sensitive modern concertgoer attending performances of these two operas might very well respond to them in approximately the same way; yet the intentions of the composers, and the sensibilities of the audiences for whom they wrote, are vastly different. This interpretive dislocation has been remarked upon in the visual arts far more often than it has in music; it is just this phenomenon to which Christoph von Dohnányi was referring.

The composers of the twentieth century, that is to say most of the best of them, have broken with the exotic and passionate musical idiom of Romanticism as cleanly as the Romantics broke from the refined, stately music of the Age of Enlightenment. Paradoxically, modern concertgoers brought up on a steady diet of Beethoven, Mendelssohn, and Brahms (and, as we have noted, Bach and Handel, who until recently were made to sound as much like them as possible) often find themselves perplexed when they hear the music of their own time. They have been conditioned to expect, and thus they demand, a thrilling emotional impact, what might be called the Romantic buzz, from music. When they hear the music of their contemporaries they are puzzled, because it does not sound like music from the last century. That might seem to contradict Christoph von Dohnányi's comment, but what he was driving at is that a sensitive modern concertgoer is more likely to arrive at a true, intuitive understanding of music composed in his lifetime, however enigmatic it might be, than of Beethoven's works; yet a simplistic misunderstanding of Beethoven is nonetheless the most common interpretive attitude of all.

One reason new music sounds the way it does is because many twentieth-century composers have returned, in their own eccentric ways, to the great theme that dominated music until the aberrant irruption of Romanticism. The avant-garde recitals and operas now being performed in Brooklyn and Berkeley and Berlin may have more in common, conceptually if not musically, with what was going on in Greece three thousand years ago than with what was happening in fin-de-siècle Vienna. The works of Bach and even most of Mozart, of post-1911 Schoenberg and Philip Glass—indeed of virtually all Western music outside the century that spans the Romantic movement—are not really about the concerns of this world. They are about the cosmic harmony that their composers believed constituted the universe.

It is only now, at the close of the twentieth century, that

Romanticism is loosening its powerful grip on the musical life of the Western world. Throughout America and Europe there is now a strong and continually growing interest in early music on the one hand, and on the other a resurgent interest on the part of contemporary composers in the cosmic theme that was the dominant strain in music throughout most of Western history.

Music contains in its essence a mystery: everyone agrees that it communicates, but how? When a poet is happy, the reader knows it because the poet has told him so; and furthermore, through the symbolism of language, the poet can explain precisely how happy he is, which delicate shades of the emotion he is experiencing at the moment, and why. Yet when we listen to the allegro of a Mozart symphony, if the performance is vivid and heartfelt, it actually creates in us the sensation of joy. It is true that music is a form of symbolic language, but it is of an entirely different species than the symbolism of language. The symbolism of language evokes external reality, however far-fetched the subjective imagery it uses to accomplish that end, while what is created by the symbolism of musical staff notation exists only in the world of ideas. (We may set aside the occasional phenomenon of programmatic music, such as Renaissance madrigals that duplicate birdsong, or the tempest in Beethoven's *Pastoral* Symphony, which conveys the idea of a storm by having the orchestra imitate the sounds of one: its very novelty consists in the non-musicality of its method.)

Somehow, Mozart's symphony, rather than telling us *about* joy, creates joy. The music *is* a zone of joy. How is that possible? The Greeks knew the answer: music and the human soul are both aspects of the eternal. The one stimulates the other powerfully and, one might almost say, with scientific precision, thanks to the essential kinship of the two. Nowadays, many people squirm when you talk to them about their eternal soul, so we use other words. Of an evening at the opera, if the music was beautifully performed, we say, "It was sublime, a transcendent experience." These words have become empty figures of speech, but they arise from the deep-seated human need to feel a connection with the Absolute, to transcend the phenomenal world. Even as science and music abandoned their classic mission—the former leaving behind the larger questions that had launched its great intellectual adventure, while the latter was turning its focus inward, concentrating on the emotional life of man rather than on the vast scheme of which it was always believed that he formed a part—the questions continued to be asked. There would seem to be an inextinguishable yearning in the human soul, almost its defining characteristic, to form these connections, to find a meaningful order in the bewildering complexity of the perceptible universe.

In this century the classics have slipped to the periphery of the curriculum, and in the place of enquiring humanism we now have condescending nihilism: the modern intelligentsia smiles at Christian fundamentalists, at credulous followers of absurd schools of psychotherapy, at adherents of what is called the New Age. Yet if people are driven to feel a connection with the Absolute by wearing crystal jewelry and listening to voices from beyond the grave, as naïve as those beliefs may be perhaps we ought not castigate them for abandoning science—for has not science abandoned them? Is it reasonable to expect that the man in the street will be content with being told, "Your life is pointless, and you are destined to be a sterile, meaningless speck of stardust, but be of good cheer: science will tell you how to power your automobile with pig droppings"?

Likewise, as glorious as the music of the Romantic era is, the anomalous nature of its purely emotional appeal ensured that it would not be able indefinitely to satisfy the needs of the audience. The world order of music is now splintering: the audience for stately performances of Romantic music, blithely unaware of how paradoxical a state of affairs that is, is slowly shrinking. It is literally dying, as the generation that grew up on the musical equivalent of the Harvard Classics, big boxed sets of "The World's 100 Great Classics," side by side with the leather-bound Shakespeare, is dying. Yet as the hegemony of Romanticism declines, the modern audience often finds itself, bizarrely, alienated from the music of its own time.

Yet despite the odds, the ancient tradition of the musical cosmos, embracing and unifying noble rationalism and ecstatic mysticism, has survived. What we may call the great theme—the belief that the cosmos is a sublimely harmonious system guided by a Supreme Intelligence, and that man has a place preordained and eternal in that system—runs throughout Western civilization, even if during the declining era of Romanticism it is a muted leitmotif. As the orthodox culture focussed its attention earthward and selfward, the impulse to connect with the universal became more and more esoteric. On the one hand it has been channelled into the compositions of the twentieth-century avant-garde, as we shall see in the later chapters of this book, and on the other hand the great theme per se has survived in what might be called the folk culture of the occult underground, now completely beyond the pale of respectable intellectual activity.

Yet until we understand the sublime cosmic order that Beethoven and his progeny overturned, to revert to our previous example, he will remain as remote from us as the ancien régime itself. The history that follows does not pretend to be a comprehensive treatment of one of the most basic and widely held systems of knowledge and belief in the history of Western culture. Yet it does aspire to be something more than simply a miscellaneous catalogue of images of the cosmic. Perhaps it is not of the utmost importance to prefix too precise an advertisement. For now, let us simply call it an anecdotal history of the symphony of science and its counterpoint, the wisdom of music, traced across the centuries from its inception up to the most bewildering period faced by any historian—the present.

↬ TWO ↬

Pythagoras, the Master

It is scarcely an overstatement to say that most of what you will read in the first chapters of this history was known to every well-educated person from the earliest days of Greek civilization until the end of the last century. Here is not the place to bemoan the decline of classical education, but it is a fact of modern life that the core of what constituted education and civilization throughout the whole sweep of Western thought is now a scholarly specialty, and a rather exotic one at that. Modern students are able to obtain degrees, even in the humanities, from the country's best colleges having read no more of the classics than translations of the *Republic*, one of Homer's epics, and the *Aeneid*. However, in order to make sense of Western music from any period, it is essential to understand its humanistic basis, which is firmly grounded in the classics.

Music and science begin at the same point, where civilization itself begins, and standing at the source is the quasi-mythical figure of Pythagoras. Although most people today know little more about him than the geometrical theorem that bears his name, Pythagoras's contribution to what we call civilization is fundamental. Arthur Koestler used a musical metaphor to describe it:

> The sixth century scene evokes the image of an orchestra expectantly tuning up, each player absorbed in his own instrument only, deaf to the caterwaulings of the others. Then there is a dramatic silence, the conductor enters the stage, raps three times with his baton, and harmony emerges from the chaos. The maestro is Pythagoras of Samos, whose influence on the ideas, and thereby on the destiny, of the human race was probably greater than that of any single man before or after him.

However, primarily because none of his works survives, Pythagoras's achievement languishes in obscurity. Everything we know about the man and his philosophy comes down to us secondhand, through the writings of his followers and the commentaries of later philosophers. Modern scholarship is deeply divided on the question of the extent to which the Pythagoras venerated by Western humanists may be identified with a historical individual, born on the island of Samos in the sixth century B.C. Furthermore, the Pythagorean tradition itself holds that the Master, as he was known to his devotees, travelled widely in Egypt, Mesopotamia, and Persia when he was a young man, and there is reason to think that he picked up many of the important concepts attributed to him on these wanderings.

Yet it is also plausible that the historical Pythagoras actually did many of the things that legend claims he did, and that he himself made all the great discoveries traditionally credited to him. There is simply no proof either way: the first documentary evidence comes long after the fact, and it is heavily prejudiced by the cult that had sprung up around the legendary Master. Many modern historians, usually for modern reasons of their own, have an overpowering need to deflate the reputations of the Great Men of History, especially those whose reputations are not buttressed by the testimony of hard facts—and most particularly those who carry the triple liability of being white, male, and European (though Samos was considered by geographers classical and modern to be a part of Asia).

We shall never know exactly to what extent the historical Pythagoras corresponds to the Master of humanist tradition, any more than we shall ever know who actually wrote the *Iliad* and the *Odyssey*, or whether all those spiritual utterances in the Gospels were really said by the historical Jesus of Nazareth. Yet for the purposes of the historian it really does not matter, for the doubts themselves are anachronistic: the point is that through the vast span of history in which Pythagorean humanism (and the study of Homer, and Christianity) were vibrant intellectual forces, there was never a shadow of doubt as to the authenticity and veracity of the tradition. You can put quotation marks around "Pythagoras" if that will make you feel more up-to-date, but it will not alter the meaning; the people who pursued Pythagoreanism over the course of thousands of years did believe, implicitly, in the historicity of the Master. The real Jesus may have been a charlatan, but the Jesus worshipped by millions of people changed the course of history; and it would be wrong, regardless of what a thousand copiously annotated doctoral dissertations may say, to attribute the *Iliad* to Anonymous.

Therefore, let us assert that Pythagoras, in or out of quotation marks, was a figure of fundamental importance to Western thought. The very word "philosophy" was coined by him; his primitive precursors were known as σόφοι (*sophoi*), the wise, but Pythagoras called himself φιλόσοφος (*philosophos*), a lover of wisdom. Nonetheless, it may not be quite accurate to characterize Pythagoras as simply a thinker or a philosopher; while he propounded a scientific view of the cosmos he was also a mystic who advocated a way of life. If *pundit* and *guru* were words not so much sullied by specious usage in recent years, we might aptly apply them to him. The Pythagorean system was not only the first attempt to answer the great questions—What is the universe? From what was it created? What is man? What is the nature of human knowledge?—it also created an overarching view of the cosmos, and showed man his place in this great scheme. Perhaps of even greater importance in establishing philosophy as a thriving enterprise in the emerging civilization, Pythagoras told his adherents how to live. The language used by his tightly knit band of followers was often ambiguous and even baffling, but it covered every aspect of life.

Even if we accept Pythagoras as a historical figure, there are some basic difficulties in approaching him. Whereas we are able to perceive Socrates and Plato, who came only a little more than a century later, as fellow creatures, men with human limitations and appetites, Pythagoras eludes every effort to encompass him, to fix him in terms that we moderns can understand. To his contemporaries he was a sage and prophet, occupying a position closer to that of Jesus or Mohammed than that of a mere teacher. Pythagoras defies categorization: a primary thinker in philosophy, mathematics, music, and cosmology, he may in fact be best thought of as one who challenges the legitimacy of categories. Anyone who conceives of Pythagoras as the inventor of a geometric theorem, the formulator of laws of music theory, and the utterer of cryptic aphorisms will miss the essence of his thought entirely, for the whole point of what he taught is the interrelatedness of all human knowledge.

Although none of Pythagoras's writings (if he in fact wrote anything) have survived, we nonetheless have a very full if not entirely trustworthy knowledge of his philosophy from the works of his prolific disciples. Just as it is an impossible task to separate the historical Pythagoras from the fabulous, semi-divine sage, in the same way we cannot distinguish with much conviction between the words of the Master and the elaborations of his followers. It is not simply Pythagoras's antiquity that is responsible for this uncertainty; the greatest bar to a reliable biography and synopsis of his thought is the secrecy he imposed upon his followers. Pythagoras, the source of so much else, is also the father of the esoteric tradition. Porphyry, the pagan apologist and follower of the Neoplatonic philosopher Plotinus, wrote a brief and entertaining biography of Pythagoras at the end of the third century A.D. In his lament for a lack of authoritative sources upon which to base his work, Porphyry betrays a contemporary note of frustration at the secretiveness of the Brotherhood (quoted here in an infelicitous early twentieth-century translation by Kenneth Sylvan Guthrie):

> When the Pythagoreans died, with them also died their knowledge, which till then they had kept secret, except for a few obscure things which were commonly repeated by those who did not understand them. Pythagoras himself left no book; but some little sparks of his philosophy, obscure and difficult to grasp, were preserved by the few people who were preserved by being scattered, like Lysis and Archippus. The Pythagoreans now avoided human society, being lonely, saddened, and dispersed. Fearing nevertheless that among men the name of philosophy would be entirely extinguished, and that therefore the gods would be angry with them, they made abstracts and commentaries. Each man made his own collection of written authorities and his own memories, leaving them wherever he happened to die, charging their wives, sons, and daughters to preserve them within their families. This mandate of transmission within each family was obeyed for a long time.

Porphyry's biography compounds the hodgepodge of fact and legend typical of the Pythagoreans with a charming overlay of sententious, homespun wisdom. In chapter thirty-four, for example, the biographer recounts the great man's favorite recipe, a mixture of poppy seed and sesame, the skin of a sea onion, daffodil blossoms, and mallow leaves, all mashed into a paste with chick peas and barley and then sweetened and stuck together with honey. Elsewhere, we learn that beans were taboo because they produce flatulence, and because of a bizarre perception that they have a particular affinity, a sort of blood kinship, with man. At moments Porphyry's Pythagoras comes across as a classical Polonius, full of kindly, trite advice. Each night before his disciples retired, Pythagoras asked them to think back over the day to see if they had left some duty undone; upon rising, good Pythagoreans began their day by pondering what they hoped to accomplish by sundown.

Moreover, the picture of Pythagoras that emerges from the surviving sources places an undue emphasis on his penchant for the enigmatic. Porphyry and the other faithful biographers love to quote Pythagoras's cryptic aphorisms and then explain what they mean—mostly, it would seem, to stress their own privileged understanding of the Master's wisdom. Here are a few Pythagorean maxims, with their glosses: "Eat not the heart" signified that one ought not afflict oneself with sorrow. "Receive not swallows into your house" meant not to admit under one's roof garrulous and intemperate people. "Do not carry the images of the gods in rings" was translated to mean that one should not be too quick about expressing one's opinions about the gods to the vulgar. There are hundreds of such admonitory aphorisms, which are sometimes called the Golden Verses of Pythagoras, and we should not pay too much attention to them. While Pythagoras may well have said many of them, the overall impression they create, that of a moralistic, riddling smart aleck, is far from the mark.

The biographical outlines of the historical Pythagoras are few and quickly sketched: according to most accounts of his life, Pythagoras was the son of a gem engraver named Mnesarchus, born on the island of Samos, in the Aegean Sea near the coast of Asia Minor. Porphyry, however, tells us that one learned source claims him for the Syrian city of Tyre. Yet on the subject of Pythagoras's education all the ancient biographers agree: he studied geometry in Egypt, where he was the first foreigner to be initiated into the mysteries of Egyptian religion; in Phoenicia he learned about "numbers and proportions"; he received his instruction in astronomy from the Chaldeans, who were acknowledged by all antiquity to be the masters of that science. Porphyry concludes, "Other secrets concerning the course of life he received and learned from the Magi." Another early biographer claims that while Pythagoras was in Chaldea he studied with Zoroaster, "by whom he was purified from the pollutions of his past life, and taught about the things from which a virtuous man ought to be free. Likewise he heard lectures about Nature and about the principles of wholes. It was from his stay among these foreigners that Pythagoras acquired the greater part of his wisdom."

The claim that Pythagoras was a disciple of Zoroaster is an interesting one. It is possible, though the life of Zoroaster is if anything even more shadowy and mysterious than that of Pythagoras. If it were true it would explain the profound congruences between the philosophical systems of the two men.

Although I have been free and loose in my references to Pythagoras as the source of the Western tradition, it is questionable how apt the East-versus-West distinction is when applied to a native of the coast of Asia Minor with intellectual roots in Chaldea. We in the West automatically take the accomplishments of the earliest Asian civilizations as belonging to "us," though in fact the Fertile Crescent, in modern Iraq, is closer to the Indus, site of ancient Indian civilization, than it is to the Aegean. Indeed, there is good reason to suppose that in earliest antiquity there existed an intellectual continuum stretching throughout Asia, even into China, and that the wall between East and West was erected at a later date. There are fundamental similarities between many aspects of the early civilizations of China and Greece, music theory not the least of them; and many attempts have been made over the years to prove that Pythagoras's contribution consisted solely of passing along the Oriental musical tradition that he picked up in Chaldea.

Conditions on Samos under its oppressive tyrant, Polycrates, did not conduce to the life of a philosopher, so Pythagoras and his family emigrated to the westernmost edge of the Greek world, the colonies of Magna Graecia in southern Italy. Here is Porphyry's description of Pythagoras at this time: "His presence was that of a free man, tall, graceful in speech, in gesture, and in everything else." He was well received in Italy, and everywhere he went he attracted thousands of followers with his passionate eloquence. So persuasive was he that in the Sicilian city of Centoripae, the tyrant Simicus, after hearing Pythagoras speak in favor of liberty, abdicated his throne and divided his property among the citizens. Around 530 B.C. Pythagoras settled at Croton, on the instep of the Italian boot, where he established his academy, the Pythagorean Brotherhood. His fame circulated throughout Magna Graecia, where he exerted considerable political influence, making him the prototype of the philosopher-king, some one hundred and fifty years before Plato invented the concept.

Pythagoras's political career ultimately undid him. At the end of his life he was the victim of a political conspiracy, and he and his followers were banished from Croton. According to one source, he died in Metapontum, about a hundred miles up the coast, in 497 B.C. Each of the ancient biographers provides a different account of the Master's death, but fire plays a prominent part in all of them. After Pythagoras's death, his disciples were persecuted and forced into exile, which is what Porphyry was referring to when he described them as being "lonely, saddened, and dispersed." However, the wisdom of the Brotherhood was conserved by the handful of followers who survived by scattering far abroad. These disciples, principally Lysis, Archippus, and an important pupil named Philolaus, eventually returned to southern Italy, where they continued to preach the Master's message. Thus the Pythagorean tradition endured and became the basis of Platonism—and of the pioneering scientific theorizing by the likes of Ptolemy and Archimedes—in the century that followed.

The Pythagorean philosophy, like Zoroastrianism, Taoism, and every early system of higher thought, is based upon the concept of dualism. Pythagoras constructed a table of opposites from which he was able to derive every concept needed for a philosophy of the phenomenal world. As reconstructed by Aristotle in his *Metaphysics*, the table contains ten dualities (ten being a particularly important number in the Pythagorean system, as we shall see):

Limited	Unlimited
Odd	Even
One	Many
Right	Left
Male	Female
Rest	Motion
Straight	Curved
Light	Dark
Good	Bad
Square	Oblong

Of these dualities, the first is the most important; all the others may be seen as different aspects of this fundamental dichotomy. To establish a rational and consistent relationship between the limited (man, finite time, and so forth) and the unlimited (the cosmos, eternity, etc.) is not only the aim of Pythagoras's system but the central aim of all Western philosophy.

What was innovative about the Pythagorean system was that it expressed these basic concepts with numbers. To understand what that means we must broaden our understanding of the meaning of numbers to include properties that most people nowadays would be more likely to consider poetic or even superstitious—in other words, to accommodate numerology legitimately alongside mathematics.

One, for example, was not simply the first number used in counting; it also, says Porphyry, denoted "Unity, Identity, Equality, the purpose of friendship, sympathy, and conservation of the Universe, which results from persistence in Sameness. For unity in the details harmonizes all the parts of a whole, as by the participation of the First Cause." Two was one plus one, but it was also the dyad, the principle of dichotomy, the mutability of everything that is; three was emblematic of things that have a beginning, a middle, and an end; four was the number of points required to construct a pyramid, the simplest of the perfect solids. The first four whole numbers were used to form the tetractys, which was among the most important Pythagorean symbols:

X
X X
X X X
X X X X

Add together these four numbers and you get ten, the perfect number that is the basis of Pythagoras's, and therefore our own, mathematics. Contained within the tetractys was the mystery of how finite form—the pyramid, which is suggested by the figure of the tetractys—emerges from the infinitude of one, the single, perfect point. The spiritual progress made by Pythagorean initiates in their journey through the mysteries of the Brotherhood is symbolized by the progression in the tetractys from the unity of one to the unity of ten. One oath of the Brotherhood went: "I swear by the discoverer of the tetractys, which is the spring of all our wisdom, the perennial fount and root of nature."

Another symbolic figure invented by Pythagoras is the

γνώμων (*gnōmōn*) (literally, a carpenter's rule), which allowed him to derive all numbers, odd and even, from one and two:

```
x  x  x  x              x  x  x  x  x
x  x  x│x              x  x  x  x│x
x  x│x│x              x  x  x│x│x
x│x│x│x              x  x│x│x│x

1, 3, 5, 7, etc.        2, 4, 6, 8, etc.
```

The odd *gnōmōn*, generated from one, is square, while that of the even numbers is oblong, just as one would expect from the table of opposites.

Aristotle, in his *Metaphysics*, presents a clear picture of Pythagorean thought, showing how it derived its vision of the physical world from pure principles of number, and, most important, how these principles were manifested by music:

> The Pythagoreans, as they are called, devoted themselves to mathematics; they were the first to advance this study, and having been brought up in it they thought its principles were the principles of all things. Since of these principles numbers are by nature the first, and in numbers they seemed to see many resemblances to the things that exist and come into being; ... since, again, they say that the attributes and ratios of the musical scales were expressible in numbers; since, then, all other things seemed in their whole nature to be modelled after numbers, and numbers seemed to be the first things in the whole of nature, they supposed the elements of numbers to be the elements of all things, and the whole heaven to be a musical scale and a number.

Here, in our first encounter with the concept of the musical universe, it is clear that the Pythagoreans did not simply discern congruities among number and music and the cosmos: they identified them. Music *was* number, and the cosmos *was* music.

Pythagoras distinguished three sorts of music in his philosophy: to use the nomenclature of a later era, *musica instrumentalis*, the ordinary music made by plucking the lyre, blowing the pipe, and so forth; *musica humana*, the continuous but unheard music made by each human organism, especially the harmonious (or inharmonious) resonance between the soul and the body; and *musica mundana*, the music made by the cosmos itself, which would come to be known as the music of the spheres.

To a modern person, the most salient observation to be made about these three classes of music is their enormous discrepancy in scale. Yet for the Pythagoreans, again, there existed among them an essential identity: a piper and the cosmos might sound the same note. That is because to Pythagoras it was purely a matter of mathematics. There was no more of a difference among these three classes of music than there was among a triangle traced in the palm of the hand, a triangle formed by the walls of a building, and a triangle described by three stars: "triangleness" is an eternal idea, and all expressions of it are essentially the same.

The laws of music were of paramount importance, for they governed the whole scope of the perceptible and even the imperceptible universe. Pythagoras considered himself above all a healer, and he used music as a remedy for every manner of sickness. Since *musica instrumentalis* and *musica humana* were of the same essence, manifestations of the same truth, then by plucking the strings of a lyre one could arouse sympathetic vibrations in the human instrument. Porphyry tells us that Pythagoras "soothed the passions of the soul and body by rhythms, songs, and incantations. These he adapted and applied to his friends." Throughout the Greek tradition, music was very literally ascribed the power to soothe a savage breast.

Classical literature abounds with anecdotes like the following, paraphrased from the biography of Pythagoras by Iamblichus, a student of Porphyry's:

A young man from Taormina had been up all night partying with friends and listening to songs in the Phrygian mode, a key well known for its ability to incite violence. When the aggravated lad saw the girl he loved sneaking away in the wee hours of the morning from the home of his rival, he determined to go burn her house down. Pythagoras happened to be out late himself, star-gazing, and he walked in on this violent scene. He convinced the piper to change his tune from the Phrygian mode to a song in spondees, a tranquilizing meter. The young man's madness instantly cooled, and he was restored to reason. Although he had stupidly insulted the great philosopher just hours before, he now addressed him mildly and went home in an orderly fashion.

The Pythagoreans, when they retired for the night, cleansed their minds of "the noises and perturbations to which they had been exposed during the day by certain odes and hymns, which produced tranquil sleep and few, but good, dreams." When they woke up in the morning, they got the day off to a proper start by playing songs that dispersed the torpor of sleep. Iamblichus concludes, "Sometimes the passions of the soul and certain diseases were, as they said, genuinely lured by enchantments, by musical sounds alone, without words. This is indeed probably the origin of the general use of this word *epode*, or enchantment."

Pythagoras's most enduring contribution to music theory was his discovery of the arithmetical relationships between the harmonic intervals. Modern scholars tell us that the famous tale of how he made this discovery is probably an ancient Middle Eastern folk tale, but the story has a homely ring and no trace of the folkish about it. Here is Iamblichus's version of the story:

Once as [Pythagoras] was intently considering music, and reasoning with himself whether it would be possible to devise some instrumental assistance to the sense of hearing, so as to systematize it, as sight is made precise by the compass, rule, and telescope,* or touch is made reckonable by balance and measures—so thinking of these things Pythagoras happened to pass by a brazier's shop, where he heard the hammers beating out a piece of iron on an anvil, producing sounds that harmonized, except one. But he recognized in these sounds the concord of the octave, the fifth, and the fourth. He saw that the sound between the fourth and the fifth, taken by itself, was a dissonance, and yet completed the greater sound among them.

Perhaps we should take a moment at this point to explain what Iamblichus means by the fourth, the fifth, and the octave. The concept of the intervals is extremely important in music theory, and basic to an understanding of its historical development.

There are two principal types of musical scales, the major and the chromatic. The best way to illustrate them is to visualize a section of a piano keyboard:

The major scale consists of seven notes, which in English are most often designated by letters. The C-major scale, thus, may be described by the white keys: C D E F G A B (or, in the system of Guido, which is still used in many parts of the world, *do re mi fa sol la si*). The other sort of scale is the chromatic scale,

---

* The Greek is διόπτρας, a Jacob's staff, an ancient surveying instrument used for measuring heights. The translator, Thomas Taylor (1758–1835), a fervent Platonist, has blithely attributed to Iamblichus knowledge of the telescope more than a thousand years before it was invented.

which includes in addition to the white keys all the black keys, known as the accidentals—that is, the sharps and flats. After the seven notes of the major scale, or the twelve notes of the chromatic scale, the whole arrangement repeats itself again and again, into the infinite and inaudible height and depth of musical sound, a fact that, again, may be confirmed by looking at the piano keyboard.

Scales consist of notes ascending (or descending) in a pattern of intervals. In the case of the chromatic scale, it is a straight progression of half steps, the smallest unit of musical interval in regular use in Western music. The major scale is an irregular combination of half steps and whole steps, which, as you may have supposed, are equal to two half steps. (They are also known as semitones and tones.) The sequence runs as follows, where WS denotes a whole step, and HS a half step: WS WS HS WS WS WS HS. If you play all the white notes of a piano in order, beginning with C (the first note on the keyboard figure above), you will find that it sounds musical and "right." There are many other scales, of course. The minor scale, for example, consists of notes arranged thus: WS HS WS WS HS WS WS. If you play the white notes of a piano beginning with A, you will be playing the natural minor scale, also known as the Aeolian scale, from the assumption that it resembles the Aeolian mode of ancient Greek music. The major and minor scales, consisting of seven notes, are known collectively as the diatonic scale, as opposed to the twelve-note chromatic scale.

Fourths and fifths are simply larger intervals within the scale. They are supremely important to Pythagoras, and indeed to anyone who thinks about music, because they are harmonious. If, beginning with C, the first note of the natural major scale, you count forward five steps, you arrive at G—the fifth, or dominant. Play C and G together and they will sound more pleasingly harmonious than any other combination of two notes on the scale. Add in the third (that is, E) and you will have a C-major chord. The octave, as you might guess from the Latin etymology of the word, is the eighth note of a major scale. If you begin with C and count up to eight, you arrive at the next C; these two notes are perfectly harmonious because they are the same. That sounds paradoxical—how can two different notes be the same? Yet there is no doubt that they are, as you will discover if you play an octave on a piano (or a lyre, or a hurdy-gurdy, or a cathedral organ).

Simply put, what Pythagoras discovered that day he walked by the brazier's shop was that there was a very exact correspondence between the abstract world of musical sounds and the abstract world of numbers. Inquisitive soul that he was, he went into the shop to enquire how it was that different hammers produce harmonious sounds when striking the same piece of iron. When he compared the hammers, he made a great discovery, which transformed music forever from the realm of happy accident to a science: the musical intervals produced by the hammers were exactly equivalent to the ratios between the hammers' weights. In other words, the six-pound hammer and the twelve-pound hammer, having a ratio of 1:2, produced a perfect octave. The eight-pound hammer and the twelve-pound hammer, having a ratio of 2:3, produced a major fifth interval; and the nine-pound hammer and the twelve-pounder, with a 3:4 ratio, produced a perfect fourth. It was of tremendous importance to the initiates of the Brotherhood that these three most basic musical intervals—1:2, 2:3, and 3:4—are all expressed directly in the figure of the tetractys. The enigmatic last sentence in Iamblichus's story, which refers to "the sound between the fourth and the fifth," is simply the recognition of the fact that those two intervals are one whole step apart, an interval expressible by the ratio 9:8, and that two notes one whole step apart are dissonant.

Pythagoras, so the story goes, went home to experiment. He hung a wooden stake from the beams of his house, and tied to it several equal lengths of gut-string. To the ends of these cords he attached weights, which reproduced the same phenomenon he had observed in the blacksmith's shop: where two weights had a ratio of 2:1, their gut-strings when plucked produced an octave; when the weights were in a 3:2 relationship they produced a fifth, and so forth. Two thousand years later in Florence, Vincenzo Galilei, the father of the astronomer, would prove that there is something not quite right in the story at this point, but we shall deal with that at the proper time.

Pythagoras then proved his discovery once more upon an instrument called the monochord, which, tradition holds, he invented. The monochord consists of a single stretched gut-string with a moveable bridge, which enables philosophers and students of music theory to demonstrate the harmonic laws. A monochord looks like this:

By moving the bridge you shorten the part of the string being plucked, just as a guitar player or a violinist produces a musical tone by shortening the string with his fingertip. If you put the bridge exactly in the middle, you will have a mathematical ratio of 1:2 (that is, the ratio of the half to the whole); when plucked, the monochord will sound a perfect octave. If you set the moveable bridge at three-fifths of the length of the string, creating a ratio of 3:2, the tone produced will be a perfect fifth—and so on.

The Greek word for ratio is λόγος (*logos*), which also means word, thought, and reason. To the early Christian philosophers it had a special significance in light of the first verse of the Gospel of John: "In the beginning was the Word [*logos*], and the Word was with God, and the Word was God." John is referring to Jesus, of course, but his meaning is very close to Pythagorean idealism, and this mystical equation was of crucial importance in rationalizing Christian acceptance of the teachings of pagan philosophy. Clement of Alexandria and the other early church fathers who wrote on music could argue that Pythagoras's identification of ratio, or *logos*, with the divine principle of universal order harmonized with the gospel's identification of *logos* with God, of which Jesus was the manifestation.

Pythagoras's discovery of the arithmetical basis of the musical intervals was not just the beginning of music theory; it was the beginning of science. For the first time, man discovered that universal truths could be explained through systematic investigation and the use of symbols such as mathematics. Once that window was opened, the light spread across the whole breadth of human curiosity—not least in the field of cosmogony. The genius of Pythagoras lay in the comprehensive way he joined the inner man and the cosmos.

Before Pythagoras, the picture of the cosmos was much closer to poetry than to science. Thales, proverbially the first Greek philosopher (who was yet enough of a scientist to predict an eclipse, in 585 B.C.), followed Homer in the belief that the world was a round island floating on the cosmic ocean. This vision was consistent with Thales's first principle, that the cosmos consisted of water. Anaximander, a younger associate of Thales and like him a citizen of the Ionian city of Miletus, introduced the conceptual element of cosmic space. His image of the earth was a cylindrical column floating in air; according to Plutarch, Anaximander's cylinder was three times as long as it was wide. It was upright and stationary, not falling or wavering, for it was the center of the newly conceived three-dimensional universe. The heavenly spheres—which make their first appearance with Anaximander—are wrapped around the earth-column, nested like layers of bark around a tree. The visible heavenly orbs are not bodies at all but rather holes in these revolving spheres that admit glimpses of a cosmic fire burning between them. Thus the stars are pinpricks in one of these dark, spherical screens; the sun is the largest of these glimpses of fire, and its perceived motion is the result of the revolution of the sphere that contains the solar hole. Anaximenes, another Milesian philosopher, reverted to Thales's notion of the flat world, which in his version floated on a layer of air. His notable contribution was the conception of the stars as ornaments attached to a crystal sphere that revolved around the earth.

As various as these images are, what they have in common is exactly that they are images and nothing more. Thales and his colleagues were asking all the right questions, but they were spinning the answers out of their heads. Why should Anaximander's column be three times as long as it was wide? Why not ten times, or ten million? Why a cylinder? The point is not to belittle the contributions of these men, for after all they set many of the basic terms, but rather to point out that they were working intuitively, not within the framework of an orderly scientific system.

When we come to Pythagoras we are in the company of a sophisticated, complicated mind, motivated by a firm notion of the importance of "getting it right." Having made his wonderful discovery of the mathematical basis for the musical intervals, he came to the conclusion that these mathematical truths must underlie the very principles of the universe. Pythagoras, who had inherited the notion of the spheres, made the logical assumption that they must make sounds in their revolutions; and, that being the case, these sounds would of necessity be musical and harmonious. The Pythagoreans conceived of the cosmos as a vast lyre, with crystal spheres in the place of strings.

The classic account of Pythagoras's vision of the cosmos comes, again, from Aristotle, in his treatise *On the Heavens*, where he tells us that the Pythagoreans believed that

> the motion of bodies of that size must produce a noise, since on our earth the motion of bodies far inferior in size and speed of movement has that effect. Also, when the sun and the moon, they say, and all the stars, so great in number

and in size, are moving with so rapid a motion, how should they not produce a sound immensely great? Starting from this argument, and the observation that their speeds, as measured by their distances, are in the same ratios as musical concordances, they assert that the sound given forth by the circular movement of the stars is a harmony.

Nowhere will you find a more concise expression of the great theme of the musical universe, which endured virtually unchanged in its essentials for more than two thousand years.

With Pythagoras, for the first time, the earth itself is pictured as a sphere. A minor writer named Alexander Polyhistor, writing in the first century B.C., provides us with a synopsis of the Pythagorean view of the universe which, although oversimplified and repetitive, makes the important connection between mathematical principles and the cosmic order:

> The first principle of things is the One. From the One came an Indefinite Two, as matter for the One, which is cause. From the One and the Indefinite Two came numbers; and from numbers, points; from points, lines; from lines, plane figures; from plane figures, solid figures; from solid figures, sensible bodies. The elements of these are four: fire, water, earth, air; these change and are wholly transformed, and out of them comes to be a cosmos, animate, intelligent, spherical, embracing the central earth, which is itself spherical and inhabited round about.

The interesting variation in Alexander's version is the distinction made between One, the first principle, and the Indefinite Two, clearly taken from the key entry of the table of opposites, Limited versus Unlimited. The idea seems to be that the first principle, which is limited, takes the unlimited (indefinite) two, and uses it as a sort of raw material; the collision between limitedness and unlimitedness breeds number in its infinitude, and from that Pythagoras makes a similar progression toward complexity, sensibility, and, finally, harmony.

The elder Pliny, the indefatigable Roman encyclopaedist to whom we owe so much of our knowledge of classical science (when we can believe him), tells us in detail exactly how Pythagoras conceived of the music of the spheres. Counting outward from the earth to the outermost sphere of the fixed stars, Pythagoras fixed the musical intervals as follows: from the earth to the moon was a whole step; from the moon to Mercury, a half step; Mercury to Venus, another half step; from Venus to the sun was a minor third, which is equal to three half steps; the sun to Mars, a whole step; Mars to Jupiter, a half step; Jupiter to Saturn, a half step; and from Saturn to the sphere of the fixed stars, another minor third. Thus the musical scale to which Aristotle referred, the so-called Pythagorean Scale, runs: C, D, E-flat, E, G, A, B-flat, B, D.* If you play it on a keyboard, it will not sound very musical; moreover, it comprises fourteen semitones, and fourteen is not a number that would have excited much enthusiasm among Pythagoreans.

To deal with the problem that none of us is able to hear these musical sounds of surpassing loudness, Aristotle explains that "the sound is in our ears from the very moment of birth and is thus indistinguishable from its contrary silence, since sound and silence are discriminated by mutual contrast." An analogy was drawn with the coppersmith, who spends all of his time surrounded by the din of his shop, until finally he grows oblivious to it. Only Pythagoras, according to the tradition of the Brotherhood, could hear and comprehend that celestial harmony.

* Arthur Koestler got this wrong in *The Sleepwalkers*; you will end up banging your head against the wall if you try to reconcile his account of the musical intervals of the spheres with the Pythagorean Scale.

THREE

Plato and the World Soul

To move from Pythagoras to Plato is to emerge from the poetic mists of esotericism to the bright, steady light of a well-reasoned public argument. Plato had an overwhelming influence, establishing the methodology and working outline of Western philosophy for the next two thousand years. The Socratic dialogue, so called because of the endearing and superbly rational presence throughout Plato's works of Socrates, Plato's master and alter ego, became the theoretical center of philosophy, its method and very soul.

Plato's relentless rationalism, doggedly worrying every question that came into its purview until a defensible truth had been arrived at, was the very antithesis of the hermetical mysticism of Pythagoreanism. Which is not to say that Plato does not have his mystical side: he assuredly does. But unlike Pythagoras, Plato is mystical without being enigmatic. When he expresses an abstruse concept, he does so as lucidly as he can— even if the reader finds himself feeling like a slow learner in one of the dialogues, pestering Socrates to spell out his ideas more explicitly.

Of all Plato's works the most mystical, and perhaps the

most mystifying, is the *Timaeus*, which is also the dialogue most explicitly in the Pythagorean tradition, for it is Plato's own exposition of the great theme. The *Timaeus* is more ambitious even than the *Republic*, for it encompasses the creation of the cosmos and all that it contains, including the human organism and psyche, from the raw stuff of what Plato calls the World Soul. The dialogue also contains as a sort of preamble, quite superfluously, the first known description of the myth, if that is what it is, of Atlantis. More likely it is a purely literary invention of Plato's, yet why he introduced into one of his most serious and ambitious works this tall tale of a great island in the Atlantic Ocean peopled by a race of brave warriors, which perished in a cataclysmic earthquake, remains quite as mysterious as the story itself.

In the Middle Ages and Renaissance, the *Timaeus* was one of Plato's most widely read works, but in the modern era it has been almost completely neglected. One reason is that it is, frankly, rather heavy going. But the principal reason might be that, by any modern standard, it is fanciful and simply "wrong." The *Republic* and the *Symposium* are the two works of Plato's most widely read nowadays, for they are considered to be still relevant to modern life: the former deals with political science and the latter with love. Yet the *Republic* and the *Symposium* are not in any sense less "wrong" or less fanciful by modern standards than is the cosmogonic view expressed in the *Timaeus*. Perhaps the decreased popularity of the *Timaeus* may be attributed to a cause already postulated, that during the Romantic era the discourse shifted from cosmic issues, once the staple of intellectual enquiry, to man. Therefore, if a modern reader is going to bend his mind into a classical pattern, in the post-Romantic era he is far more likely to do so in order to learn what the Greeks thought about the state, or about love, than he is to waste his time on a lot of superstitious rubbish about the World Soul.

That attitude, of course, is narrow-minded and unfair, for the *Timaeus* is one of the most beautiful expositions of Greek thought to survive. Its real business is to explain why what is is, and why it is the way it is. It contains what is still one of the most persuasive answers to that most compelling of questions—Why is there a universe at all?—which may be summarized briefly. First, the world is a fair place, characterized by goodness and order. Second, such a place cannot have come into being except through some cause, and that first cause, Plato's Demiurge, must have been himself good and orderly. Third, since that which is good is entirely lacking in envy, the Demiurge wanted his creation to be as good and as orderly as himself, and so he created the world in his own image.

Today such an explanation has a Panglossian flavor; we have become accustomed to the cynical notion that the world is very far from being fair and orderly. In order to understand Plato, we must try to think of the universe as it was known throughout most of the history of Western civilization, before cynicism acquired respectability. That the cosmos was not only orderly but in fact epitomized and actually defined order was a fundamental principle that would not be questioned for another two thousand years. Plato could not have believed Dr. Pangloss's maxim that all is for the best in this best of all possible worlds. But he did believe, firmly, that the reason it was not was no reflection on the cosmos itself, but was rather due to the imperfect understanding of man. Plato would have condemned the suggestion that the universe was a random, fearful place as the warped fancy of a diseased mind, only thinkable by someone who had been eating the wrong foods, reading the wrong books, and listening to the wrong music.

Something of a hybrid between a myth and an exercise in logic, the *Timaeus* has baffled many of the best minds that have attempted to penetrate its mysteries. Even Cicero, who translated it into Latin, wrote that he "never could understand that mysterious dialogue." The dialogue's Demiurge is portrayed as a sort of math-crazed tailor, measuring out swatches and snippets of the World Soul in a phenomenological crazy quilt that satisfies logical and mathematical imperatives more than it corresponds to any recognizable image of the perceptible universe. The eight pages (in translation) of the *Timaeus* that Plato devotes to the creation of the cosmos have generated thousands of pages of commentary, yet no one has ever quite managed to clarify Plato's obscurities satisfactorily. What the dialogue does communicate, unambiguously, is Pythagoras's final triumph. The cosmogonic vision of the *Timaeus* is the mystical Pythagorean equivalence of music, the cosmos, and mathematics brought out of the esoteric closet and thrown open for inspection by all thinking persons.

Once Plato has established the motive for the first cause, and its rightness, he then sets forth the basic terms of the argument through his mouthpiece, Timaeus. First, he establishes that there is only one universe (the etymology of the English word from *unus* makes that much clear). Speaking pure Pythagorese, Timaeus asserts, "Have we, then, been right to call it one Heaven, or would it have been true rather to speak of many and indeed of an infinite number? One we must call it, if we are to hold that it was made according to its pattern. For that which embraces all the intelligible living creatures that there are, cannot be one of a pair." This universe is composed of the four elements familiar to primitive Greek philosophy: earth, air, fire, and water. The universe, in the beginning, had the form of a sphere: "And for shape he [the Demiurge] gave it that which is fitting and akin to its nature.... He turned its shape rounded and spherical, equidistant every way from center to extremity—a figure the most perfect and uniform of all; for he judged uniformity to be immeasurably better than its opposite."

Although Plato begins his description of creation with this account of the contrivance of the physical universe, he immediately assures us that the Demiurge had already created the soul before he made the body. Accordingly, he backtracks a bit and explains to us how the soul was created. The First Cause compounded the soul of three elements: Existence, Sameness, and Difference. Up to this point, everything has been readily predictable from Pythagoras's table of opposites, but now it begins to get a bit complicated, and a few Platonic ground rules, unstated in the *Timaeus*, need to be explained. In an earlier dialogue, the *Sophist*, Plato made a distinction between existence and sameness, which, he explained, is analogous to the two different meanings of "is": the word can mean that something *exists*, and also that something is *the same as*. F. M. Cornford, the scholar whose commentaries are an invaluable starting point in understanding the *Timaeus*, illuminates the problem by showing that you can say of anything (1) that it exists; (2) that it is the same as itself; and (3) that it is different from some other thing. These three properties are the constituent parts of the soul.

The Demiurge first mixed together Difference, "hard as it was to mingle," with Sameness, and then he added in Existence, which resulted in a malleable substance, the World Soul. The Creator thoroughly interpenetrated the World Soul with and wrapped it round the spherical universe. From this raw material he was now ready to create the cosmos. At this point, emboldened by Pythagoras, Plato soars off into the world of pure number:

> And having made a unity of the three, again he divided the whole into as many parts as was fitting, each part being a blend of Sameness, Difference, and Existence. And he began the division in this way. First he took one portion from the whole [equals 1], and next a portion double of this [equals 2], the third half as much again as the second, and three times the first [3], the fourth double of the second [4], the fifth three times the third [9], the sixth eight times the first [equals 8; 9 precedes 8 because it is a square number, while 8 is the cube of 2], and the seventh twenty-seven times the first [27].

What are we to make of this? Is Plato simply grabbing numbers from the air? The reader may be wondering how anyone could possibly say, as I have done, that Plato is not being enigmatic. Yet he is trying to be as lucid as he can: he never promised us that the creation of the cosmos was going to be an easy business. Now we must turn to the commentators for help. Crantor, a philosopher from Asia Minor writing just decades after the death of Plato, constructed a diagram based upon the Greek letter lambda to help us to visualize the relationships of these numbers:

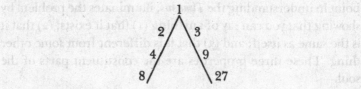

The principal insight offered us by this little schema is that it relates Plato's numerical system to the Pythagorean table of opposites. On the left side are the even numbers, representing the unlimited; on the right are the odds, signifying the limited. The choice of 1 and 2 and their sum, with their squares and cubes, is not hard to justify, but somewhat more elusive is the question of why Plato's Demiurge stops with 27. The reason for stopping with the cube, according to Cornford, is that the cube symbolizes body in three dimensions, all that are needed to create a universe. Coincidentally, by ending his progression at 27, Plato brought the total number of integers to seven, which will serve his purposes later on.

The important point about Plato's numerical division of the World Soul is that it is explicitly musical, and thus firmly in the Pythagorean line. His initial set of numbers may be expressed, in the key of C major, on a modern stave like this:

1   2   3   4   8   9   27

The Demiurge proceeds to fill in the intervals between these notes by two means, the harmonic and the arithmetic, which Plato defines in terms hopelessly confusing to anyone brought up on modern mathematics. Yet, always trying to be as clear as he can, Plato spells it out for us: "These links gave rise to intervals of 3:2 and 4:3 and 9:8 within the original intervals"— intervals equivalent, it will be recalled, to the fifth, the fourth, and the whole step, respectively. If we fill in these intervals between the tones already posited (and eliminate duplicates), we arrive at this sequence:

1 $\frac{4}{3}$ $\frac{3}{2}$ 2 $\frac{8}{3}$ 3 4 $\frac{9}{2}$ $\frac{16}{3}$ 6 8 9 $\frac{27}{2}$ 18 27

We are almost there. "And he went on," says Timaeus, "to fill up all the intervals of 4:3 with the interval 9:8, leaving over in each a fraction . . . with the numerical proportion of 256:243." That last, improbable-looking ratio is very close to the mathematical expression of a half step. By interpolating this final batch of intervals, the Demiurge arrives at an almost perfect major scale covering five octaves. The following figure, adapted from Cornford, is added for the sake of completeness, to show (approximately) how the addition of these last intervals in the first octave fills in all the gaps of the seven-note scale:

That finishes the business of constructing the mathematical underpinnings of the World Soul; and, happily, "by this time the mixture from which he was cutting off these portions was all used up."

It ought to be emphasized that there is no element of *musica instrumentalis* in Plato's exposition. It would never have occurred to Plato or any of his students to play a few bars of the World Soul on the lyre: it was a purely theoretical system. Nor did the principle of the harmony of the spheres enter into his cosmogony. That beautiful, quintessentially Pythagorean concept will come into play in the *Republic*, as we shall see, in Plato's Myth of Er. But for the purpose of diagramming the World Soul, he confines himself to pure, unprettified figures.

Having divided the World Soul into all these musical intervals, the Demiurge takes on the task of physically constructing the heavens. The World Soul is now conceived of as a long, malleable band. (The transition is very abrupt, and the reader is left to his own devices to see how this cosmic band relates to the spherical universe, interpenetrated with and wrapped round by the World Soul—perhaps an implicit recognition that Timaeus is spinning elaborate metaphors more than reliably reporting cosmic events.) This soul-band is slit lengthways into two long strips, which are joined in the middle and bent round to form two rings. These represent the sidereal equator, the fixed band of stars around the horizon, and the zodiac, which is variable.

As he wrote the *Timaeus* Plato very likely had before him an armillary sphere, the elaborate schematic representation of the cosmos. An armillary sphere comprises a number of nested spheres, cut away into sections or rings similar to those described by Plato, so that the viewer can see their interrelationships. Typically, the outermost of these rings is fixed, representing the sidereal equator, which contains what the Greeks called the fixed stars, while inner rings represent the variable stars, the zone of the zodiac (which, in the *Timaeus*, the Demiurge subdivides into the planets, each of which will occupy its own celestial sphere).

The two rings of the cosmos are identified with a Pythagorean dichotomy: the outer, stationary ring is the Same, while the inner, variable ring is the Different. The Same, being as it were from the positive side of the table of opposites, moves to the right, the "good" direction; while the Different, playing yin to the yang of the Same, moves to the left. Continuing to follow the table of opposites, the Demiurge ordained that the Same would remain One and indivisible, while its companion he divided into many—seven, to be exact: "The inner revolution he split in six places into seven unequal circles, severally corresponding with the double and triple intervals, of each of which there were three." Concerning this section, Cornford states, "At this point there is some obscurity about the procedure of the Demiurge." One might be more inclined to call it irritating vagueness: What, exactly, does Plato mean by "severally corresponding" (in the Greek, ἐν λόγῳ δὲ φερομένους), which another translation renders, not more helpfully, as "related proportionately"? After having gone to such great pains to subdivide the World Soul into a lovely and harmonious system, Plato now seems to be retreating behind the cover of fuzzy language.

The explanation for this is not a failure of the Platonic nerve; the great man never hesitated to go out on a limb. Rather, it is likely that he was being intentionally vague. The great debate about the motion of the planets was already well under way by the time Plato was writing, and he was scientist enough to recognize that the celestial observations that had been made at that time were not yet sufficient to discover the

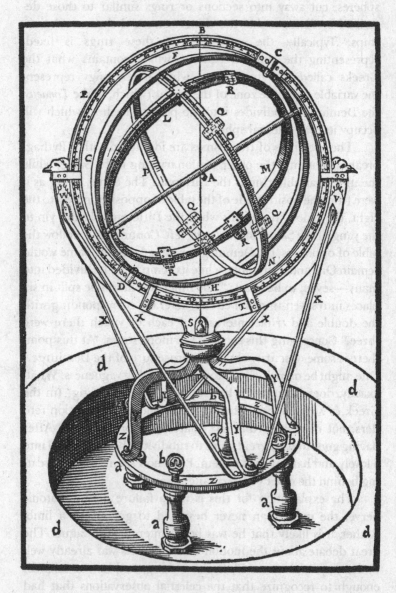

Armillary Sphere, from Tycho Brahe's *Astronomiae instauratae mechanica*

underlying logic of planetary motion. That such a logic existed he would never have doubted, and he would likewise have been secure in the belief that the celestial logic, once it was understood, would be reconcilable with a sublime system of mathematical harmony. Yet if there was one principle that Plato taught virtually as a commandment, it was that man's understanding of the cosmos was faulty and incomplete, and the corollary of that principle was that it was far better to concede these inadequacies than to blunder forward arrogantly and perhaps impiously.

Given that, it is significant that Plato is willing to go so far as to say, first, that the number of cosmic rings (which may be thought of as being synonymous with the Pythagorean spheres) is equivalent to the number of notes in the musical octave, and second, that the intervals between them "correspond severally" with (or "relate proportionately" to) the mathematical intervals from which he has just derived the diatonic scale, even if he is unable to be more specific about those correspondences. The former point is especially telling, for in order to bring the total to seven, the number of tones in the scale and the number of the variable stars—the five known planets plus the sun and the moon (or, to make it eight, for a full octave, we may include the indivisible ring of the Same, which may be taken as representing the outermost sphere of the fixed stars)—Plato had to omit any mention of the earth. According to Aristotle's unsympathetic account, the Pythagoreans invented a new planet, something called the counter-earth, simply in order to bring the total to ten, a perfect number (the seven variable stars, the fixed stars, earth, and the spurious counter-earth make ten). Yet we may conclude, in a general way, that while Plato was not secure in the details, he nonetheless subscribed without reservation to the Pythagorean concept of a musical universe based upon mathematical principles of harmony.

Plato reveals his true feelings about astronomy in the eighth book of the *Republic*, when Socrates explains that concentrating on the movements of the visible stars is to miss the point: a wise man seeks to apprehend the pure forms that underlie the heavenly bodies. "These intricate traceries in the sky are, no doubt, the loveliest and most perfect of material things, but still part of the visible world, and therefore they fall far short of the true realities—the real relative velocities, in the world of pure number and all perfect geometrical figures, of the movements which carry round the bodies involved in them. These, you will agree, can be conceived by reason and thought, not seen by the eye." Plato thus provides the astronomer a wonderful way out of any difficulty: if the stars do not move according to one's theory, then the fault lies either with the observer, whose perception is muddled, or with the stars themselves, since they, too, are only gross and unreliable reflections of the perfect forms that lie beyond our ken.

The account I have given of the cosmic creation described in the *Timaeus* is far from complete: Plato goes on to describe the motion of the planetary rings in an ambiguous fashion that has kept the pots of commentators boiling for centuries. He then entangles himself in a fine mess over the question of whether time is dependent upon celestial motion, and asks us to consider whether the motion of the planets might not be used to calculate time. Before the dialogue is finished, he has taken on the entire range of observable phenomena in the physical world, from solid geometry to the common cold.

However, before we leave the *Timaeus*, we ought to mention how Pythagoras explains the harmony of octaves. In his section on circular motions, he explains that musical sounds are like little darts propelled by the musician, when, for example, he plucks the strings of his lyre. These darts, which have a circular motion, come into contact with the circular motion within the human organism—the *musica humana*. When these two motions are in concord, we experience the pleasurable sensation of acoustic harmony. Timaeus concludes, "So they give a thrill to fools and true enjoyment to the wise, by reproducing divine melody in mortal movements."

As we have seen, Plato the philosopher, unable to chart with any precision the celestial spheres, left them imprecise. Where Plato the philosopher left off, Plato the literary artist took over. His modus operandi seems to have been something like: When you cannot establish the truth with a certainty, write a beautiful myth. One of the most beautiful he ever wrote was the Myth of Er, the final passage of the *Republic*, an evocative firsthand tour of the celestial spheres.

Er was a soldier slain in battle who was permitted by the gods to return from the afterlife to tell humanity what awaits us on the other side of death. First, Er tells us, he journeyed with the souls of the other dead soldiers to a marvelous place called the Meadow, where there were a pair of openings in the earth and a pair of openings in the sky. In between sat the Judges. There was a constant stream of souls coming and going in this place: those newly arrived were judged, and sent either into the sky, if they had been just, or down into the earth, if they had been unjust. The term of reward or punishment was a thousand years. Meanwhile, through the other openings, the souls that had just finished their thousand-year terms were returning from the sky or from the underworld, in order to return to a new life on earth.

These souls were given a seven-day respite in the Meadow, and on the eighth day they were sent on their way. After a five-day journey, they arrived at a "straight shaft of light, like a pillar, stretching from above throughout heaven and earth, more like a rainbow than anything else, but brighter and purer." This is the Spindle of Necessity. (Notice that the number of days these celestial voyagers spend in the Meadow is seven, the number of notes in the diatonic scale; add to that the number of days they must travel to reach the Spindle of Necessity and you have twelve, the number of notes in the chromatic scale.) A spindle is a primitive form of spinning wheel, a wooden dowel tapering at its ends which is fitted with a whorl, a doughnut-shaped weight, commonly made of stone. To make yarn the spinner pulls the wool or other raw stuff through the tight nexus between the spindle and the whorl.

The Spindle of Necessity, Er tells us, was made of adamant, while the whorl was made "partly of adamant and partly of other substances." The cosmic whorl is described as a set of nested bowls, the top surfaces of which create a series of rings. "For there were in all eight whorls, set one within another, with their rims showing above as circles and making up the continuous surface of a single whorl round the shaft, which pierces right through the center of the eighth." The eight circles of the whorl, beginning with the outermost and moving toward the spindle, are: the ring of fixed stars, which is spangled; the moon; the sun, which is brightest; Aphrodite (Venus); Hermes (Mercury); Ares (Mars), "somewhat ruddy"; Zeus (Jupiter), which is whitest; and Kronos (Saturn), similar to Hermes and somewhat yellower than the others. The names of the planets are given in the *Epinomis*, a dialogue doubtfully ascribed to Plato, where it is suggested that the Greeks took the names for the planets from the Syrians, substituting the names of Greek gods for their Syrian counterparts. Of course, our own names for the planets are those of the equivalent Roman gods.

On each of the rings there stood a Siren, and as the celestial whorl revolved, the Sirens sang, each one a different note "so that all the eight made up the concords of a single scale." Here also were seated the Fates, the daughters of Necessity, who greeted the cosmic voyagers and allowed them to choose their new lives. "It was indeed, said Er, a sight worth seeing, how the souls severally chose their lives—a sight to move to pity and laughter and astonishment." He saw souls of the just foolishly pick unjust lives, and chastened unjust souls picking just ones; men choosing new lives as women; beasts taking human shape and vice versa. After the voyagers had chosen, they drank from the forgetful waters of Lethe. "At midnight there was thunder and an earthquake, and in a moment they were carried up this way and that, to their birth, like shooting stars"—all except Er, who was permitted to return to the earth to tell what he had seen. He concludes his tale with a plea for justice:

> If you will believe with me that the soul is immortal and able to endure all good and ill, we shall keep always to the upward way and in all things pursue justice with the help of wisdom. Then we shall be at peace with Heaven and with ourselves, both during our sojourn here and when, like victors in the Games collecting prizes from their friends, we receive the prize of justice; and so, not here only, but in the journey of a thousand years of which I have told you, we shall fare well.

In his edition of the *Republic*, which I have quoted here, F. M. Cornford advances the intriguing notion that the Myth of Er shares some narrative features with works on the same theme by Empedocles and Pindar, and that all of them may be based upon a very early ceremonial work from the Orphic tradition— the *Ur*-Er, as it were. Orphism, the oldest form of mystery religion in the Greek world, is centered around the veneration of the Thracian poet Orpheus, who rescued his beloved Eurydice from Hades, god of the underworld, by the enchanting strains of his lyre. Cornford suggests that the Myth of Er and the other orphically inspired literary works may be taken directly from dramatic representations of *tableaux vivants* which were shown to initiate worshippers into the Orphic and other mystery religions. As late as the fourth century A.D., when paganism was in its death throes, one of the most popular cults, that of Mithras, still used a similar myth of the voyage into the underworld as the centerpiece of its initiation rite. The neoplatonist Celsus, writing in the second century, asserts that the Mithraic rites helped the soul to make its ascent through the heavenly spheres.

The story of Orpheus unites two key elements of Er's story,

and indeed of the Pythagorean concept of the harmonious universe: the power of music and the renewal of life. Cornford enumerates some specific ways in which the Myth of Er seems to echo these hypothetical Orphic myths of harmony and renewal. The term of a thousand years for the souls of the dead, for example, is based upon the idea that the reward or punishment is tenfold, the perfect Pythagorean figure; starting with the figure of one hundred years for the span of life, another Pythagorean number, and increasing it tenfold yields one thousand.

In the eighteenth century, there was a great rebirth of interest in Orphism among German scholars. J. M. Gesner, a classicist at the University of Göttingen, wrote an essay contending that there must have been a historical Orpheus, a great poet and musician who established a secret fellowship that would serve as the pattern for the Pythagorean Brotherhood and all the other mystic cults to follow. According to Gesner, it makes no more sense to doubt the actual existence of Orpheus than that of Pythagoras or of Saint Francis of Assisi.

One concept that Plato treats at some length in the *Republic*, which is certainly derived from the legend (or, as Gesner would have it, the works) of Orpheus—is the capacity of music to heal, and hence its importance in education. We have already seen how these Orphic powers were attributed to Pythagoras, in the story about the lad from Taormina. Plato is even more specific about the spiritual effects of the different modes, which were the different configurations of notes for tuning the lyre. Here is not the place to enter into a detailed treatment either of the modes of Greek music, which are imperfectly understood (at best) and horrifically complicated, or of Plato's musical ethics. However, the importance Plato places upon music in education is of crucial importance, since it was taken for granted throughout the whole history of the West that music was a defining human activity, and therefore every educated person was trained in the rudiments of music.

All Greek writers on music dwell at length on the modes, which may be approximately likened to the keys of modern music theory. It is a time-honored doctrine in Western composition that a cheering piece of music must be written in a major key, while a melancholy composition must be in a minor key. That notion is no longer quite so prevalent as it once was, but it nonetheless persists in some circles, a vestige of the quasi-magical powers that the Greeks attributed to the modes. In the *Republic* Socrates tells us that the Mixolydian and the hyper-Lydian modes are dirgelike and ought to be done away with, for they are useless "even to women." The Ionian and certain Lydian modes, on the other hand, are relaxing and convivial; because of their softening influence he would prohibit youths from singing or hearing them. That leaves the Dorian and Phrygian modes, which he allows: the former because it emboldens warriors and helps them accept and cope with setbacks, the latter because it has potent persuasive powers to induce temperance, moderation, and law-abidingness. The reader will recall, if Socrates does not, that it was the Phrygian mode which inflamed the jealous boy from Taormina to madness; one can only speculate whether such inconsistencies were apparent to Plato's readers. Of course, it is possible that the Phrygian mode meant something different in sixth-century Sicily than it did in fourth-century Athens; it is also possible that Iamblichus, writing seven hundred years after Plato, may have muddled his modes.

Having dealt with melody, Plato now takes on rhythm. First he lays down the law that music must fit speech, not the other way round. (So far as we know, cultivated Greeks did not listen to purely instrumental music, but rather considered it to be only a vehicle for song—that is, poetry—or dance.) Plato makes it explicit that the speech to be imitated in rhythmic compositions is that of a man whose life is brave and orderly; the particular rhythms that will convey these qualities, however, he leaves to Damon, the musical authority of Athens, to determine. Damon, who figures in other Platonic dialogues, is known to have believed that strictly preserving the conventions of musical styles was fundamental to maintaining political and social order.

Plato agrees with him, and explains why: "The decisive importance of education in poetry and music is this: rhythm and harmony sink deep into the recesses of the soul and take the strongest hold there, bringing that grace of body and mind which is only to be found in one who is brought up the right way.... Approving all that is lovely, he will welcome it home with a joy into his soul and, nourished thereby, grow into a man of a noble spirit."

As Plato develops his theme, he clarifies what he means by those terms, and especially what he does not mean by them. To understand the Platonic ideal of education, we must free our minds of the modern aestheticism most of us take for granted. "Ars gratia artis" (art for art's sake) is a Romantic coinage of the nineteenth century; for Plato, such a tag would have been incomprehensible, if not anathema. "Ars gratia republicae" would have been, for him, the only conceivable formula. What makes a thing beautiful is not that it induces a pleasurable sensation per se. Beauty is that which ennobles and leads a person to a just and temperate life.

Thus, for example, there ought to be nothing sensual in the joy experienced by youths raised to be just men and to appreciate what is beautiful. In the dialogue, Glaucon presses Socrates on this point, finally making him concede that he considers music that is "pan-harmonic"—that mixes every variety of rhythm promiscuously—to be quite as dangerous for a youth as rich Attic pastries or licentious sexual relations. Averroës, the great Arab philosopher, wrote a commentary on the *Republic* (known to us only from a Hebrew translation) that explicates this with unequivocal candor:

For there is nothing at all in common between a sound mind and pleasure. That is because pleasure throws a sharp-minded man into a perplexity resembling a madman's, all the more when he goes to excess. For example, the pleasure of copulation, more than anything else, will drive a man mad. Hence, pleasure should not be mixed with the desire of the musical one; rather he should desire only the beautiful with self-control. . . . This is the end at which the activity of music aims.

This highly moralistic view of music seems extreme to us, though the early Christian fathers found it quite a congenial point of view. Almost in the same breath, Plato tells us that lovers should only be permitted to kiss each other in the way a father kisses a son, and that any physical contact more intimate than that ought to be strictly prohibited. A person who transgressed this limit would be stigmatized as showing "want of taste and true musical culture." That stretches a modern reader's credulity too far: anyone who has read classical drama or history knows that the ancients were not more successful at repressing the appetites of the flesh than were the Victorians or any other moralistic society, and that copulation and other purely pleasurable activities enjoyed a scandalous popularity then as now.

Indeed, one might be tempted to ask, How seriously are we meant to take this? The answer, I think, is, very seriously indeed: Plato is dealing in ethical principles, not probabilities, and the fact that vicious intemperance flourished in his society would not have been a serious argument against inculcating high ideals in the minds of the young. Yet for the present purpose, the important point, setting aside all the ethical considerations, is that for Plato, and thus for the Western intellectual tradition that was to follow, music was the key to the human soul, the most potent instrument available to man for enlightenment.

❦ FOUR ❧

"The Key to the Universe"

Once it took root in the philosophy of Plato, the Pythagorean conception of the universe as a musical-numerical system rapidly became the standard throughout the Mediterranean world. While there were refinements and variations, some more successful than others, the basic terms remained those established by the Master. Aristotle, not content with a cosmos that made elegant mathematical sense, tried to create a mechanical model of the universe that would actually work. He postulated a complicated series of buffer spheres between the planetary spheres in an attempt to explain the seemingly random and disconnected movements of the stars. He ended up with no fewer than fifty-five spheres, each motivated by its own spirit—its Intelligence—to account for the motions of the seven planets. As Arthur Koestler says, "It was an ingenious system—and completely mad, even by contemporary standards." Despite Aristotle's enormous prestige, his fifty-five spheres were soon forgotten.

Yet the most egregious blunder in ancient astronomy was the shifting of the center of the universe from the sun to the earth, a notion that received its most influential expression in the works of the second-century A.D. astronomer Ptolemy. Like Aristotle before him, Ptolemy devoted a huge amount of time and ingenuity to an attempt to account for the perceived variability in planetary motion. To this end he invented a baffling system of epicycles, pirouetting mini-revolutions executed by the stars in the course of their circular orbits. It was a disastrous detour in the history of astronomy, for the Ptolemaic idea of a geocentric universe remained the official cosmology for some fourteen hundred years, at first in the academy and then in the church, until Copernicus rediscovered heliocentricity (a doctrine first propounded in the astronomical theories of the Pythagoreans).

With Ptolemy, however, we are straying from our course: the stargazer of Alexandria might profitably be thought of as the first "techie." His great work, the *Almagest*, was nothing more than a compendium of astronomical observations—immensely useful to navigators, certainly, but thin gruel to someone interested in pondering the larger questions of science. The numerical universe of Pythagoras and Plato's Demiurge were never of any practical use, but were rather intended to help man find his place in the universe. And that, of course, was exactly what the Greeks thought the real purpose and meaning of science to be.

Yet among those thinkers who still addressed the basic questions of science, the Pythagorean view had triumphed utterly by the first century B.C. The best proof we have of this is to be found in the works of Cicero. Although the great humanist had written that he never could understand the *Timaeus*, he was nonetheless sufficiently moved by it to translate portions of it into Latin. One of Cicero's most popular works—and, according to some modern classicists, a key masterpiece of Latin prose—is a fable called *Scipio's Dream* (*Somnium Scipionis*). Based upon Plato's Myth of Er, the *Dream* forms the conclusion of Cicero's treatise on government, *De republica*, just as Plato's myth concludes his similarly titled work about the temporal state. It is not the clearest account of the music of the spheres ever written, either from a musical or a cosmogonic point of view, but it is probably the most widely read. Because of the unparalleled fame of its author, perhaps the most influential prose writer of antiquity, and because it was composed in Latin, the scholarly tongue of all Europe, *Scipio's Dream* was read continuously from the time it was written down to the eighteenth century—when it formed the basis of an opera by Mozart, which we shall examine later.

As its title indicates, Cicero's description of the universe is presented at an epistemological remove from the phenomenal world, as a dream—and the dream of a mortal at that. For the man who coined the phrase "polite atheism" to describe the religion of the Romans was not about to leave something as important as the origin of the universe to the mercies of the gods, or even in the hands of a philosopher-mathematician Demiurge. The ultra-patriotic Cicero's natural choice for the leader of a voyage through the cosmos was the great warrior Scipio Africanus, hero of the Punic Wars and renowned defender of Roman freedom.

The dream is that of Scipio's grandson, Scipio Africanus the Younger. The tale begins with the younger Scipio recounting that once upon a time, when he was in Africa visiting the barbarian king Manissa, who had been a staunch friend and ally of his grandfather, he fell into a deep sleep, and the elder Scipio appeared to him. He recognized his grandfather from his portrait busts; he had not known him in life. The elder Scipio first makes some prophecies to the dreamer about his future glories—all perfectly accurate, of course, since Cicero is writing after the fact. Then Scipio admonishes his grandson about the supreme importance of maintaining civic justice and order. He tells him, "That you may be more zealous in safeguarding the republic, Scipio, be persuaded of this: all those who have saved, aided, or enlarged the republic have a definite place marked off in the heavens where they may enjoy a blessed existence forever."

Having established the rationale for this excursion into the afterlife—that is, to inculcate a love of civic virtue by showing the rewards that await a just man after his death—Cicero proceeds to describe a manifestly Pythagorean cosmic order, with one difference: man figures somewhat more prominently here than he has up to this point. Man, Cicero tells us, was created expressly as the guardian of the earth, with a mind that is of the same substance as the eternal fires that constitute the fixed stars and the planets, "those spherical solids which, quickened with divine minds, journey through their circuits and orbits with amazing speed." Young Scipio is stunned by the transcendental beauty of the cosmos, and he tells the reader that he is ashamed when he sees how puny is the earth in comparison with the rest of creation.

Then the old man explains the disposition of the heavens. Outermost is the sphere of the fixed stars, called the celestial sphere by Cicero, which contains the other seven spheres. These are, from the outer edge to the center: Saturn, Jupiter ("propitious and helpful to the human race"), ruddy Mars, the sun, Venus, Mercury, and lastly the moon. Beneath the moon, and according to the Ptolemaic order the center of all things, is the earth, which is stationary: there, all is transitory except for the immortal souls of men. In *Scipio's Dream*, then, the earth is the ninth sphere; it would appear that numerological imperatives had receded in importance by the time of Cicero.

Now the dreamer exclaims, dumbfounded, "What is that great and pleasing sound that fills my ears?" Scipio Africanus's reply is an admirably clear and concise exposition of the great theme:

> That is a concord of tones separated by unequal but nevertheless carefully proportioned intervals, caused by

the rapid motion of the spheres themselves. The high and low tones blended together produce different harmonies. Of course such swift motions could not be accomplished in silence and, as nature requires, the spheres at one extreme produce the low tones and at the other extreme the high tones. Consequently the outermost sphere, the star-bearer, with its swifter motion gives forth a higher-pitched tone, whereas the lunar sphere, the lowest, has the deepest tone. Of course the earth, the ninth and stationary sphere, always clings to the same position in the center of the universe. The other eight spheres, two of which move at the same speed, produce seven tones, this number being, one might almost say, the key to the universe. Skilled men, imitating this harmony on stringed instruments and in singing, have gained for themselves a return to this region, as have those who have cultivated their exceptional abilities to search for divine truths. The ears of mortals are filled with this sound, but they are unable to hear it. Indeed, hearing is the dullest of the senses: consider the people who dwell in that region called Catadupa, where the Nile comes rushing down from lofty mountains; they have lost their sense of hearing because of the loud roar. But the sound coming from the heavenly spheres revolving at very swift speeds is of course so great that human ears cannot catch it; you might as well try to stare directly at the sun, whose rays are much too strong for your eyes.

The tone of the tale is ambiguous: like Plato, Cicero wants to give us some specific information, to imbue the musical universe with a palpable, "scientific" reality, but unlike Plato he does not want to go to the bother of elaborating the mathematical rationale for it. Thus the effect of *Scipio's Dream* is primarily literary. When Cicero proclaims that seven is the key to the universe, the harmonious number of the celestial spheres (plus, as usual, the octave, the sphere of the fixed stars), he is only reporting a consensus.

Another departure from the Platonic model in *Scipio's Dream* is the emphasis on man, particularly on man the music-maker. As we have seen, in the Pythagorean-Platonic cosmos music exists quite independently of man: *musica instrumentalis* is harmonious because it reflects the perfection of the cosmos in the world of ideal forms; an octave sounds harmonious to human ears because the rhythms of the music are in concord with our own internal rhythms, the *musica humana*. Yet Cicero clearly makes the connection between excellent music and the excellent life. "Skilled men" (*homines docti*) who are able to imitate the music of the spheres on stringed instruments or with their voices are permitted to return to heaven.

Exactly what Cicero is saying here about the afterlife is unclear; the idea that skillful musicians are somehow specially privileged to an eternity of celestial bliss, which is what he seems to be saying, is preposterous. Yet another interpretation suggests itself: Cicero may be harking back to the legend of Orpheus, whose playing on the lyre had magical powers, and the entire passage may have an esoteric significance. The word *docti*, "skilled," is the past participle of the verb *doceo*, to teach or instruct; *homines docti* thus means, literally, "men who have been taught." *Docti* is generally used to mean learned or clever, particularly in the sense of being especially good at something, like music. However, it may also be taken to mean "those men who have been initiated or indoctrinated." The members of the Pythagorean Brotherhood and the adepts of the Orphic cult could certainly have been called *docti*.

Another phrase in the text that supports this esoteric interpretation is the one used to describe the number seven: Cicero tells us that it is *rerum omnium fere nodus*. I have followed the very free translation of William Harris Stahl, "the key to the universe," which captures the sense of the Latin, but the literal translation is "the knot of almost everything." While *nodus* means a knot in all the senses that we have in English—a knot in a tree, a knot tied in the hair, a knot on the forehead, and so forth—it can also mean a bond or obligation, a knotty problem, or a perplexing or enigmatic thing. It is in this last sense that

Cicero uses the word: if we could solve all the mysteries posed by the number seven, he is saying, then we would have the answer to almost everything—the key to the universe.

This passage from *Scipio's Dream* marks a turning point in the evolution of the great theme. Pythagoras and Plato completely identified mathematics, music, and philosophy: the first two are essential preparations for the last, and all three are aspects of what every enlightened person understood science to be—the state of knowing. The search for excellence in the life of the individual and the search for justice in the society were dual aspects of the search for the ultimate truths. Yet by the time of Cicero the process of disintegration had begun. The intellectual quest was splintering. If we may apply some modern attitudes to the intellectual scene in the first century A.D., we get a scenario that goes something like this:

Those scientists who were concerned with the phenomenal world, the number-crunching likes of Ptolemy, were beginning to disdain the idealistic spirituality of the primitive philosophers; numbers were needed for recording the movements of the heavenly bodies, which was a useful thing to do, and that was enough. The notion that they signified something deep and spiritual was an enigma imponderable by mortal men. Insoluble questions such as these were best left to the tortuous maunderings of the philosophers.

On the other hand, the philosophers began to disdain those whose ken did not exceed the limits of the phenomenal world. Why go to such laborious lengths to collect astronomical data, when we know quite well from Plato that all observable phenomena are only illusions, untrustworthy reflections of ideal forms which no mortal will ever be able to apprehend? We have already seen that Cicero had no interest in going into the mathematical basis for the music of the spheres. He must have studied the musical ratios in order to translate Plato's *Timaeus*, and we know from Cicero's essays that he held Pythagoras in the highest esteem, but for his purpose, which was to inspire civic virtue and a love of justice, numbers per se no longer had any usefulness. All that endless multiplying and dividing of the World Soul in the *Timaeus*, one suspects, was what Cicero was thinking of when he said that he "never could understand that mysterious dialogue"; one can easily imagine him skipping over the boring parts and going straight to the good stuff, the cosmic mythopoesis.

The foregoing sketch, admittedly, is an oversimplification. Thinkers for centuries to come certainly considered themselves to be both scientists and philosophers. The intellectual exponents of the new religions—Augustine and Averroës, to name two towering examples—would have disapproved of any suggestion to the contrary. Yet a distinction between the two had quietly slipped into existence: now there were scientists who philosophized and philosophers who applied themselves to scientific work.

What is remarkable is that music was an essential part of the curriculum of both camps. Music remained an important constituent of mathematics in European education until the nineteenth century; and the spiritual aspect of *musica instrumentalis*, of ordinary, mortal music making, began to take on an ever more important role. The whole notion of individual expressiveness begins if only obliquely with Cicero's *homines docti*, and very slowly builds to an overwhelming crescendo in the music of the Romantics, where the voice of the individual is of supreme importance.

Before we leave *Scipio's Dream*, let us pause long enough to gloss Cicero's ingenious proof of immortality. The elder Scipio Africanus tells his dreaming grandson that the outward form of a man is not his true self; that honor must go to his mind, which is godlike, because of its sovereign power to propel the body. Anything that is able to move itself is immortal—for whereas anything that is not self-moving must rely upon an outside agency for life, a self-mover never ceases to move. If a self-moving entity (such as man) had a beginning, that would be tantamount to saying that it had to rely upon an outside force for motion, to propel it into being, which is clearly impossible, for in that case it would not be self-moving at all. Therefore the soul must always have had the motivating power within itself; what has always existed must also be eternal and without the possibility of ending. Since it is clear that anything that is self-moving is eternal, who would deny that that describes the condition of the soul?

Young Scipio is persuaded. His grandfather tells him therefore to focus his energies on worthwhile pursuits, which in the Ciceronian scheme means above all else to serve Rome, so that he will be able to return to the realm of the celestial spheres for his reward. Then the old man departs abruptly, and the dreamer awakes.

If Plato established Pythagoras's vision of the musical cosmos in the mainstream of higher thought, with *Scipio's Dream* it became a commonplace. After Cicero one encounters the concept everywhere, in the work of hack rhetoricians and uninspired encyclopaedists as often as in the writings of great and prestigious philosophers. A few examples will suffice.

Athenaeus was an early third-century pagan rhetorician and grammarian whose principal work was a hodgepodge called *The Sophists at Dinner*, a sort of classical *Bartlett's Familiar Quotations* vaguely in the form of Plato's *Symposium*, which collects together pithy quotations from all sorts of sources. Some of it is authentic table talk, like this quotation from one Anaxilas: "Music is like Libya, which, I swear by the gods, brings forth some new creature every year!" There is also a liberal amount of pseudo-Pythagorean nonsense, such as Athenaeus's prescription for curing sciatica by playing the aulos in the Phrygian mode over the affected area. His discussion of Pythagoras is chatty and simplistic but correct in its essentials:

It is plain to me that music should also be the subject of philosophic reflection. Pythagoras of Samos, with all his great fame as a philosopher, is one of many conspicuous for having taken up music as no mere hobby; on the contrary he explains the very being of the universe as bound together by musical principles. Taking it all together, it is plain that the ancient "wisdom" of the Greeks was given over especially to music. For this reason they regarded Apollo, among the gods, and Orpheus, among the demigods, as most musical and most wise.

The Pythagorean concept of cosmic harmony was eagerly taken up by the early Christians, who found it to be readily adaptable to their own ends. Clement of Alexandria used music as a principal metaphor in his *Exhortation to the Greeks*, a treatise written in the late second century to reveal the errors of paganism and the perfect truth of Christianity. Clement, like so many of the early church fathers, had been brought up a pagan, and his writings are imbued with the convert's zeal. The *Exhortation* begins by ridiculing the Greek myths concerning music, especially the legends about Orpheus, whom he contemptuously calls the "Thracian wizard." He tells us that the songs of the pagan poets are delusions inflicted upon the singers by the evil arts of demons and sorcerers. Then he launches into a rhapsody on the saving power of Jesus, who is known in the *Exhortation* as the New Song.

"See how mighty is the New Song!" exults Clement. "It has made men out of stones and men out of wild beasts. They who were otherwise dead, who had no share in the real and true life, revived when they but heard the song." Of course Clement's game is to take all the marvelous properties of Orphic song and re-attribute them to Jesus. Clement was an educated man, and he must have known that he was transmitting, almost unaltered, the philosophy of the Master of pagan sages, Pythagoras: "Furthermore, it is this [the New Song] which composed the entire creation into melodious order, and tuned into concert the discord of the elements, that the whole universe might be in harmony with it."

Clement's image of Jesus as a song is derived from the Ninety-eighth Psalm, the source of these famous verses: "Make a joyful noise unto the Lord, all the earth: make a loud noise, and rejoice and sing praise. Sing unto the Lord with the harp; with the harp, and the voice of a psalm. With trumpets and sound of cornet make a joyful noise before the Lord, the King." In the *Exhortation*, Clement likens the human body to a musical instrument more explicitly than anyone before him; the bodies of Christians are like David's harps and trumpets, and Jesus is the musician, playing upon them a hymn of joyful thanksgiving to God the father. "What is more, this pure song, the stay of the universe and the harmony of all things, stretching from the center to the circumference and from the extremities to the center, reduced this whole to harmony, not in accordance with Thracian music [i.e., the pagan music of Orpheus], which resembles that of Jubal, but in accordance with the fatherly purpose of God, which David earnestly sought." Thus literally did Clement construe the notion of *musica humana*: Jesus, the Son of God, who was descended from David yet embodied the *logos*—pure reason and the concept of musical ratio, as well as the Word of God—and who thus existed eternally, "arranged in harmonious order this great world, yes, and the little world of man, too, body and soul together; and on this many-voiced instrument of the universe He makes music to God, and sings to [the accompaniment of] the human instrument."

According to the fourth chapter of Genesis, Jubal, son of Lamech, was the first musician, "the father of all such as handle the harp and organ." Clement nowhere explains why the music of this legendary Hebrew patriarch resembles that of the false demigod Orpheus, rather than that of King David, nor do modern commentators shed light on the subject or even take note of it. The answer may be that Jubal was descended from

Cain, whose line perished in the Deluge, rather than from Seth, progenitor of Noah, whose virtue in the eyes of the Lord saved mankind from extinction.

One of the principal tasks of the medieval scholastics was to reconcile the wisdom of the ancient philosophers, which they accepted implicitly, with the rigorous ethical demands of Christianity. A great deal of ingenuity went into this task, which sometimes approached pure syncretism, that is, simply changing the names of pagan concepts and entities to Christian ones. We have already noted the identification of the celestial musical ratios with the Word of God arising from the ambiguity of the Greek word *logos*. Another example of this process is the way the heavenly Intelligences, which motivated the cosmic spheres in Aristotle's cosmic scheme, are transformed in the Christian era into angels. An enormous amount of scholastic literature is devoted to angels—their hierarchy, their duties, their true nature—questions that remained unsettled long after they ceased to matter. Jumping forward several centuries to Thomas Aquinas, we find that the question of angels, and whether or not, and how, they turned the crystal spheres of the cosmos, was still a matter of heated debate. In an uncharacteristically succinct statement on the subject, Thomas wrote to a friend in Venice: "That the celestial bodies are moved by spiritual beings, I cannot think of one philosopher or saint who would have denied it; and if we admit that the angels move the spheres, no official source, that is, no saint could deny this proposition. . . . Therefore the conclusion is: Everything that moves in nature is moved by the Ruler; the angels transmit the motion to the spheres."

The greatest of the early Christian philosophers, Augustine, composed a lengthy treatise early in the fifth century called *De musica*, the first in a long line of scholarly, theoretical treatments of the Pythagorean theme. It is not much studied nowadays, even by Augustinian specialists. The common wisdom about the work is that it is tedious and derivative, though the erudite modern musicologist Nino Pirrota calls it "a great work." Augustine himself is partly to blame; in a letter to the Bishop of Capua he apologizes for the work's unrevised state and calls it trivial and childish.

Yet in the Middle Ages the work was enormously influential. In it what Boethius would later call the quadrivium, the "fourfold path"—the division of mathematics into arithmetic, geometry, music, and astronomy—was codified and took the form that would dominate the curriculum in Europe for fourteen centuries. The concept was not original with Augustine; the Pythagorean Archytas had first put the idea forward in the fourth century B.C., and Plato himself took it up in his formulation of the philosopher's education in the *Republic*. Yet Augustine's prestige in the Christian intellectual world, which had completely supplanted (and just as completely assimilated) the pagan intellectual tradition, made the quadrivium a fundamental part of education in the cloister just as it had been in the stoa.

Augustine formulated a definition of music that was quoted for centuries: "Music," he wrote, "is the science of modulating correctly." Not much of a definition, and such as it was he borrowed it from the Roman scholar Varro, but throughout the Middle Ages it was cited by every music theorist, major and minor. Augustine also made an important advance in fundamental musical philosophy, if that is the right way to describe it. In his *De ordine*, he delineates the essential duality of heard music:

And since what the intellect perceives (and numbers are manifestly of this class) is always of the present and is deemed immortal, while sound, since it is an impression upon the sense, flows by into the past and is imprinted upon the memory, Reason has permitted the poets to pretend, in a reasonable fable, that the Muses were the daughters of

Jove and Memory. Hence this discipline, which addresses itself to the intellect and to the sense alike, has acquired the name of Music.

According to Augustine and the scholastics who followed him, this mind-body polarity, which operates in a similar fashion in all the performing arts (after all, if the audience at the performance of a tragedy does not remember what happened in the first act, then the denouement will not make much sense), has a fundamental meaning in music. Yet it is important to keep in mind that the idea of music as a performing art does not really become established until the Renaissance. As the quotation from *De ordine* makes clear, for Augustine as much as for Pythagoras himself, pure music is number made audible. It does not exist for us humans except insofar as we are able to apprehend it through the medium of the senses, yet its meaning exists in the realm of pure intellect.

Such abstruse ideas, perforce intended for an educated elite, were propagated in cheap, simplified form by the encyclopaedists and grammarians of the late empire. Augustine's concept concerning music and memory, for example, was summarized by the Spanish encyclopaedist Isidore of Seville in cursory and inaccurate fashion: "It was fabled by the poets that the Muses were the daughters of Jove and Memory. Unless sounds are remembered by man, they perish, for they cannot be written down."

Another, more trustworthy, transmitter of the classical tradition was Cassiodorus, secretary to Theodoric, the king of the Ostrogoths. Cassiodorus wrote a big book called the *Institutiones*, an encyclopaedic synopsis of all learning, written for the instruction of monks. His pithy summary of the great theme, incorporating Augustine's definition of music, is not exactly wrong, but like that of Athenaeus it has a mechanical quality, as though he is repeating something of which he has only a superficial grasp: "Music indeed is the knowledge of apt modulation. If we live virtuously, we are constantly proved to be under its discipline, but when we commit injustice we are without music. The heavens and the earth, indeed all things in them which are directed by a higher power, share in the discipline of music, for Pythagoras attests that this universe was founded by and can be governed by music."

Undoubtedly the best of the medieval writers on music was Boethius, whose *Principles of Music* (*De institutione musica*), composed early in the fifth century, was a prestigious and widely read textbook of music theory. Boethius falls somewhere between encyclopaedist and philosopher: there is little original thinking in his work, yet he is unmatched in the clarity of his thinking. His treatise consists almost entirely of quotations from the principal Greek texts, which he translated into Latin and pasted together in a coherent order. Boethius is extremely important, for it was he who set the terms. Through him, the basic vocabulary of classical music theory is transmitted to the late Middle Ages and the Renaissance: the quadrivium is called that for the first time in Boethius; the categories of *musica mundana*, *musica humana*, and *musica instrumentalis* are first enumerated by him; and the critical distinction between *musici*, theorists of pure music, and *cantores;* mere singers and instrumental performers, is a Boethian formula.

Boethius is unflinching and plainspoken in his definition of music as an ethical science. While that may seem almost paradoxical to us moderns, we must again bear in mind that from the scholastic, as from the classical, perspective, the true and the right converge. For Boethius, there is no division between the mind and the soul. The seeker of knowledge will inevitably arrive at spiritual understanding as well:

> Thus it follows that, since there are four mathematical disciplines, the others [arithmetic, geometry, and astronomy] are concerned with the investigations of truth,

whereas music is related not only to speculation but to morality as well. For nothing is more consistent with human nature than to be soothed by sweet modes and disturbed by their opposites.... Thus we can begin to understand the apt doctrine of Plato which holds that the soul of the universe is united by a musical concord. For when we compare that which is coherently and harmoniously joined together within our own being with that which is coherently and harmoniously joined together in sound—that is, that which gives us pleasure—so we come to recognize that we ourselves are united according to this same principle of similarity.

Manuscript copies of *The Principles of Music* are to be found in a great many medieval libraries, an indication of the work's prestige in the Middle Ages. The most amazing testimony to its influence is the fact that it remained a standard text for the teaching of music theory at Oxford until 1856.

One of the most bizarre remnants of the late classical era is *The Marriage of Philology and Mercury*, an allegory by the fifth-century Carthaginian satirist Martianus Capella. The work has long since lapsed into oblivion, but throughout the end of the Roman Empire and the Middle Ages it was a highly influential summation of the liberal arts, a sort of fanciful myth *cum* encyclopaedia. In a stilted, obscure style that alternates between prose and poetry (known technically as Menippean Satire, after its originator, the Cynic Menippus), Martianus Capella's work chronicles the search for a suitable bride for Mercury, and the god's subsequent betrothal and marriage to Philology, the love of learning. Philology's bridesmaids at the wedding feast are personifications of the seven liberal arts—the quadrivium plus the trivium, which comprised grammar, dialectic, and rhetoric—each of whom makes a speech expounding the basis of her field.

These seven speeches were, as usual, constituted of bits and pieces filched from standard Greek sources, compounded with considerably less skill and understanding than is to be found in, for instance, Boethius's treatise, with which it is roughly contemporaneous. William Harris Stahl, the classicist who devoted the last years of his life to unravelling the tortuous rhetoric and convoluted conceits of Martianus's work, eventually came to despise his subject, and his commentaries are a masterpiece of withering academic nastiness. Yet despite all its flaws, *The Marriage of Philology and Mercury*, especially the speeches of the seven liberal arts, was a popular and influential textbook for centuries.

Stahl tells us that medieval readers were impressed by the "hauteur and jargon of Martianus's female wizards," and "titillated by his spicy descriptions of their charms and by other fillips which he added to the setting." That is certainly an apt characterization of Martianus's lavish description of Harmony (who comes last, in the place of honor, among her sister liberal arts), which seems uncannily to anticipate the opulence of Renaissance operatic settings. Stahl's translation of Harmony's entrance at the wedding feast is worth quoting in full:

> Harmony walked along between Phoebus and Pallas, a lofty figure, whose melodious head was adorned with ornaments of glittering gold. Her garment was stiff with incised and laminated gold, and it tinkled softly and soothingly with every measured step and movement of her body. Her radiant mother the Paphian [Venus]—who followed her closely—though she too moved with graceful measure and balanced steps, could scarcely match the gait of her daughter. In her right hand Harmony bore what appeared to be a shield, circular overall, with many inner circles, the whole interwoven with remarkable configurations. The encompassing circles of this shield were attuned to each other, and from the circular chords there poured forth a concord of all the modes. From her left hand the maiden held, suspended at equal length, several small models of theatrical instruments, wrought of gold. No lyre or lute or tetrachord appeared on that circular shield, yet the strains

coming from that strange rounded form surpassed those
of all musical instruments. As soon as she entered the hall,
a symphony swelled from the shield.

It would appear that in his description of Harmony's shield
Martianus is attempting to describe something that he has seen
but does not understand. What ingenious goldsmith's creation
can he be referring to? And what are those golden theatrical
instruments?

   After a few soothing hymns, Harmony gets down to the
business of presenting the elements of her science to the divine
assembly. Her discourse is pure Pythagoreanism, not in the
scientific mode of Boethius but rather in the mystical vein. In
many ways, Martianus's musical disquisition is more strikingly
an adumbration of Renaissance Neoplatonism than it is a true
reflection of classical thinking. In any case, it is a vital link in
the evolution of the great theme. Harmony begins by telling
the guests that from the moment she was begotten by the
Creator, "I have not forsaken numbers. I followed the courses
of the sidereal spheres and the whirling motions of the entire
mass, assigning tones to the swiftly moving celestial bodies."
That is formulaic Pythagorean stuff, but what follows is very
close to the esoteric beliefs of the Neoplatonic cults, which
would convey the essence of classical music theory through
later centuries. The "Monad" invoked by Harmony in the
following quotation is, in the first place, a Pythagorean term
for the number one, the principle of oneness, but it is also
similar to the quasi-religious usage of the word that would
later be employed by no less deep-dyed a Neoplatonist than
Giordano Bruno (from whom Leibniz would borrow the word
as the basic term of his philosophy):

   But when the Monad and first hypostasis of intellectual light was conveying to earthly habitations souls that
   emanated from their original source, I was ordered to
   descend with them to be their governess. It was I who

designated the numerical ratios of perceptible motions and
the impulses of perfect will, introducing restraint and harmony into all things. . . . I deigned to have numbers underlie the limbs of human bodies, a fact to which Aristoxenus
and Pythagoras attest.

Harmony then proceeds to describe the manifold blessings
that music has bestowed upon mankind, the familiar catalogue
of the efficacy of song in soothing the soul, curing the body,
petitioning the gods, inspiring soldiers to bravery, taming wild
animals, and the rest. Her preamble concluded, she then em-
barks on a garbled and superficial gloss of Greek musical
theory.

Music by the time of the decline and fall of the Roman
Empire had gone as far as it could go in its severely limited
technical state of development. The reason I have provided so
few details of classical music theory is because its obsolete terminology renders it all but unintelligible to the modern reader;
almost any sentence from Boethius's *Principles of Music* would
require pages of explanation. He was still using the technical
vocabulary formulated by the Pythagoreans more than a thousand years before and handed down to him virtually unchanged by Augustine and the other scholastics.

More important, the Greek system was very clumsy as a
tool for musical performers. A vast amount of the music of the
classical and medieval eras survives to us in manuscript form,
particularly from the Byzantine school, but it is seldom if ever
performed because we have no firm principles to guide us in
establishing performance practices. There was no useful way
of indicating rhythm, and the pitches specified by the rudi-
mentary Greek system are now almost entirely a matter of
conjecture.

Time had come for a change.

The Renaissance Musici

By the tenth century music was being made that might sound familiar to our ears: the polyphonic chant of the church at Cluny. The settings of the mass there, and then throughout Europe, notably at the cathedral of Notre Dame de Paris, are the first music known to us based upon principles of harmony in the sense that we use the word today, that is, the simultaneous production of concordant musical notes. For Pythagoras, Plato, and the other purveyors of the great theme, harmony was something that existed in the mind: if you heard a piper play a note, and then immediately afterward its dominant, or fifth, say C followed by G, the harmonious relationship between the two notes would be apparent. Yet since a piper or a singer can only produce one note at a time, it did not occur to ancient musical theorists to play harmonious tones simultaneously. It is quite true, of course, that a lyre or cithara can produce several tones at one stroke, and a singer and a lyre can, and do, harmonize easily. More to the point, the famous story of Pythagoras and the anvils clearly suggests that the concordant tones were being produced in unison, as does Plato's account of the Sirens in the Myth of Er. Yet it is nonetheless a fact that for the Greeks harmony meant successive, not simultaneous tones; and the notion of concerts, that is, aggregates of musicians playing in consort, is a medieval one.

With the Cluniac settings of the mass there began a flowering of polyphony that would dominate the music of Europe for several hundred years. The development of polyphony was facilitated in part by the invention of the organ, which made it possible for cathedral singers, who were often poorly schooled, to sing their parts in tune and keep together. In fact, it is quite possible that the organ actually inspired the invention of polyphonic harmony; for the instrument's capability to produce several consonant tones simultaneously might well have prompted the theoretical discovery necessary for the first composers in the new musical form. In any case the growth of polyphony and that of the new instrument are inextricably linked.

Yet the principal requirement for the development of complex music was a complex method of notation. The confusing Greek system was almost useless, which meant that composers and performers often had to rely entirely upon memory. Greek musical treatises frequently invoke the importance of memory; and while that faculty was no doubt highly developed among musicians of the classical era, the lack of a standard method for recording musical ideas was a crippling limitation on composers.

The first inkling of modern notation comes to us from Boethius, or, more properly, from ninth-century manuscripts of his *Principles of Music*, in which for the first time the letters of the Latin alphabet are used to denote the tones of the diatonic and chromatic scales, which are derived, in approved classical style, from divisions of the monochord. Yet what is lacking in the so-called Boethian notation is the key element of repeatability, a unified system which could be used for any musical purpose. This breakthrough was achieved by a monk of the abbey of Saint-Amand named Hucbald. In a treatise called *The Principles of Harmony*, which dates to the early tenth century, Hucbald makes plain that what he means by *consonantia*, the Latin word for an instance of harmony (as opposed to the concept of *harmonia*, which could of course be applied with equal correctness to music, people, or the cosmos): "*Consonantia* . . . is a prescribed and harmonious mixing together of two sounds, which will come about only if two different notes coincide at the same time in one *modulatio*, as when a man's voice and a boy's voice sing at the same time."

Hucbald goes on to show how the pitches of the notes of a composition may be indicated graphically, through a series of parallel lines that are labelled in the margin with capital T's and S's, to signify "Tone" and "Semitone." The words to be sung are actually written on the corresponding tone-line.

The last big step toward the development of standard notation was the idea of using both the lines and the spaces of the staff, an innovation that originated with a Benedictine monk named Guido, choirmaster of the cathedral of Arezzo. In a lively manual for the use of music masters, called the *Prologus in antiphonarium*, Guido sets forth in practical terms a standard method of notation that could be used to express any musical composition. Guido's names for the notes of the major scale were derived from a setting of a popular hymn to Saint John, in which the first syllable of each line named the pitch on which it was sung (for the last note of the scale, the name was obtained by abbreviation):

> *Ut* queant laxis
> *Re*sonare fibris
> *Mi*ra gestorum
> *Fa*muli tuorum
> *Sol*ve polluti
> *La*bi[i] reatum,
> *Sancte Io*hannes.

[O Saint John, in order that thy servants may be able to sing the praises of the marvels of thy deeds upon loosened harpstrings, cleanse those accused of the stain of sin.]

Guido's scale is still in use in Italy, France, and other places in exactly this form, although somewhere along the line Ut was dropped in favor of the familiar Do (except in France, where Ut is still preferred).

Guido explains his aim succinctly in the *Prologus*: "I have decided, with God's help, to write this antiphoner so that hereafter, by means of it, any intelligent and studious person may learn singing and so that, after he has thoroughly learned a part of it through a master, he will unhesitatingly understand the rest of it by himself without one." He then proceeds to give a vigorous testimonial on behalf of his system, in the huckstering style of Dulcamara in *L'elisir d'amore* (or, to choose an example from our own century, Robert Preston in the title role of Meredith Willson's *The Music Man*): "Should anyone doubt that I am telling the truth, let him come, make a trial, and see what small boys can do under our direction, boys who until now have been beaten for their gross ignorance of the psalms and vulgar letters, who often do not know how to pronounce the words and syllables of the very antiphon which, without a master, they sing correctly by themselves, something which, with God's help, any intelligent and studious person," and so forth and so on.

Guido is to be forgiven if he gets a bit carried away with himself, since the change wrought by his system, or rather by the coherent notational theory his system perfected, was truly revolutionary. The *Prologus* is the very model of a music textbook, presenting the basics of music theory and notation with a clarity few of its modern counterparts can match. Guido explains in plain and precise language the conceptual underpinnings of the system, and then propounds some clever mnemonic devices to facilitate its assimilation. In addition to the jingle method employed by the hymn to Saint John to teach the notes of the scale, he also invented an ingenious system known as the "Guidonian hand," which enabled music masters to teach the scale using the hand as a visual aid, and he perfected the use of colored lines on the staff to denote uniform pitch.

The *Prologus* was an immensely popular work and survives in many manuscript copies. Guido was so successful that he was brought to Rome, where he sold Pope John XIX on his system, and thereafter it was adopted throughout Christendom. It is no exaggeration to say that without the Guidonian method, music in the West could not have evolved in the complex form that it did. Before Guido, it was necessary for every new piece to be memorized, which severely limited the field of possibilities. Through the use of Guido's notation and the organ, richer and more complex polyphonic compositions were suddenly possible, and before long the Holy See was complaining that the words of the sacraments were no longer being taken seriously for their own sake but were rather seen merely as pretexts for musical compositions.

The earliest polyphonic compositions are known to us only through musical examples cited in manuals of music theory, which, fortunately, constitute one of the most plentiful literary genres in the late Middle Ages. All these works are purely Boethian in outlook; passages may be found in works separated by hundreds of years that are nearly identical. Here, for example, is a passage from the *Scholia enchiriadis*, an influential polyphonic treatise composed circa 900, which transmits unaltered the basics of Pythagorean-Boethian theory in reasonably clear style: "Music is fashioned wholly in the likeness of numbers. . . . Whatever is delightful in song is brought about by number through the proportioned dimensions of sounds. . . . Sounds pass quickly away, but numbers, which are obscured by the corporeal element in sounds and movements, remain." More than four hundred years later, in 1319, the Parisian mathematician and astronomer Jean de Muris wrote in a speculative treatise called *Ars novae musicae*, "Sound is generated by motion, since it belongs to the class of successive things. For this reason, while it exists when it is made, it no longer exists once it has been made.... All music, especially mensurable music, is founded in perfection, combining in itself number and sound."

Both writers are propounding essentially the same idea, but there is an important distinction to be drawn between their modes of expression. The anonymous author of the *Scholia* is simply restating, uncritically and for the thousandth time, the Pythagorean premises he had himself been taught, while Jean de Muris is beginning to approach the subject experimentally. In the *Ars novae musicae*, Jean constructs an elaborate philosophical-mathematical edifice that at first blush seems to have little to do with his stated subject, *musica practica*. He devotes many pages to a highly theoretical tribute to the perfection of the number three and all numbers divisible by three— eminently Pythagorean but not terribly useful from a musical point of view, for it would be centuries before the rhythm we know as the waltz would be used as the basis for musical composition. Then he gets down to his real business, which is to codify and integrate the many conflicting notational methods that were then current for indicating the time values of notes, a vitally important step. Jean's treatise is also notable for containing one of the first accounts of a keyboard instrument, which he describes as a monochord of nineteen strings (the prefix "mono-" obviously having lost any literal force), which was plucked by a mechanism put into play by the keys. Called a chekker, this instrument was the forerunner of the virginals.

Although Oliver Strunk, in his invaluable collection of *Source Readings in Music History*, places Jean at the end of the Middle Ages, it really makes more sense to see in him the first stirrings of the Renaissance: while Jean is still solidly based in Pythagoreanism, his outlook is enquiring and humanistic. The very title of his work has a certain "modern" ring to it: the phrase *ars nova* comes at a time when music making, and indeed all the arts, began to be infused with a quickened spirit and a heightened degree of technical innovation. The word "new" takes on special importance beginning in the fourteenth century: Dante, Jean's near-contemporary, was composing verse in what he called the *dolce stil novo*, the sweet new style; later, when Francis Bacon first expounded his understanding of the principles of the scientific method, he would call his book the *Novum Organum*. Nor is the usage an exaggeration: after centuries of lifeless recapitulation, the characterization "new" was more than justified.

In the Middle Ages, in just the same way that the *musici* were strictly confined to reiterating the principles of classical theory, heavily flavored with the conventional piety of the age, artists were restricted to creating pious works for the use of the church. Illuminated manuscripts, cathedral architecture, and sculpture and painting for the church were the only creative outlets for the visual arts in the medieval era. Then, paradoxically, the new spirit of the arts in the Renaissance found its expression in an old source: the pagan themes of classical antiquity that had for so long been interdicted by orthodoxy. The liberating spirit also moved the composers of Renaissance music, and in much the same way, to attempt to recreate the musical arts of antiquity. Yet despite all the earnest theorizing of its creators, Florentine melody of the sixteenth century no more resembles Attic music than a Botticelli Venus resembles the work of a Greek painter.

The reason why that is so marks a turning point in the history of musical science. As we have seen, the *musici*, the musical theorists of the Middle Ages, regarded themselves very much as scientists; and at the same moment that the *cantores* were finding new modes of expression, exploring profane themes and purporting to revive the musical methods of antiquity, a parallel change was taking place in the science of music. The new spirit of lively intellectual enquiry that was sweeping through the physical sciences, and indeed making them over, was also challenging the basic assumptions of classical music theory—the very essence of the great theme. It was no longer enough, in any field of intellectual activity, simply to continue restating the received wisdom of past ages. For the first time since Pythagoras himself, the basic premises of music were being subjected to rational scrutiny.

We must take care not to overstate this phenomenon: it was not as though everyone went about for centuries believing that the earth was flat and that sonorous crystal spheres in the sky were being pushed round and round by angels, and then suddenly one day sweet reason burst on the scene, dispelling superstition. It is just as true that medieval science, particularly in the Arab world, made a number of valuable contributions as it is that some of the greatest innovators of modern science were devout practitioners of alchemy, astrology, and other disciplines now considered downright anti-intellectual. The line between superstition and speculative science has always been blurry, and we may find examples as readily in our own century as any other. Is Sir Fred Hoyle any less a great astronomer for his heterodox views on extraterrestrial life? Are B. F. Skinner's contributions to modern psychology tainted by his zealous proselytizing for vitamin therapy?

Although there has been a considerable amount of Orwellian revisionism in the history of science on this point, free enquiry has always embraced the freedom to be wrong, sometimes wildly wrong. The intellectual orthodoxy of the twentieth century requires that the alchemist be presented simplistically, and fallaciously, as a witchy crackpot, like the bearded, conical-hatted sorcerer in Walt Disney's *Fantasia*; on the other hand, Kepler's extensive astrological practice and Newton's lifelong profession of alchemy are simply censored from the textbooks. As a rule, most scientists who have made great discoveries did not know that that was what they were doing at the time. While the investigative process is actually going on, great breakthroughs are frequently indistinguishable from great mistakes—just as no one sets out to write a bad symphony or a tedious opera. It takes many scientific trials, many performances, for the "truth" to out.

Yet even with this caveat, it is certain that in the musical science of the Renaissance a profound change was taking place: at the same moment that reverence for the music of antiquity was becoming the rage among composers and performers, the *musici* were for the first time applying objective criteria to the dogma of classical theory, and finding some cracks in the crystal spheres. It was actually the vogue for polyphony in the late Middle Ages which highlighted some weaknesses in the venerated Pythagorean system. So long as musical performance called for a solitary voice (whether a single performer or a chorus singing in unison) to sing or play from memory, the standards were approximate enough that no audible problems were posed by certain inherent weaknesses in the system. But when standard notation made it possible to create more complex music, some faults in the Pythagorean standard became evident. Our investigation of the great theme in the Renaissance era will begin by examining these refinements in music theory, as science began to reveal complexities and shortcomings unsuspected by the Greeks; and then in the next chapter we shall turn to the first artistic treatments of the great theme of harmony of the spheres, which led to the emergence of the opera and the ballet.

There is a problem with the musical intervals described by Pythagoras: they do not quite add up. The discrepancy, as it happens, was actually discovered by the Master himself. As much as it would have suited Pythagoras's cosmological and ethical system for it to be otherwise, he found that the ratios he discovered between the harmonious musical tones (whether at the smithy of legend or from his teachers in Egypt or Persia) varied by a very slight degree. To explain this discrepancy we must return to the vexatious question of the musical intervals; yet before we begin, I must confess that in my original presentation of the issue, I omitted to mention one important hitch.

The difficulty arises in the relationship between octaves and fifths. Among the experiments that tradition holds Pythagoras performed after his breakthrough at the brazier's shop was the following, which he carried out using two monochords. He took one of them and divided it in half seven times, thereby spanning eight perfect octaves. The other monochord he shortened by two thirds, thereby creating, as we know, a harmonic fifth. He repeated this procedure on the second monochord twelve times. Arithmetic would lead one to expect the two series to arrive at the same pitch, but in fact the tones produced by the monochords in the experiment were slightly, but perceptibly, different. The series of fifths was about one-ninth of a whole step sharper than the note produced by the octave series. This interval, mathematically expressed by the ratio 531,441:524,288, is known as the Pythagorean comma.

By the sixteenth century, the Pythagorean scale was being supplanted by a new system called just intonation, which attempted to rectify this deficiency of the traditional scale. To accomplish this it introduced the third, an interval, expressible by the mathematical ratio of 5:4, which spans three diatonic steps. For example, the third of C is E, the middle note of a C-major chord. The Pythagorean system did not recognize the third—for the excellent reason that, using the notes of its scale, the third was not harmonious. The third was not useful in the church music of the early Middle Ages, but with the evolution of polyphony this drawback of the Pythagorean scale became glaringly apparent.

Just intonation attempted to correct the problem by basing its scale upon both pure fifths and pure thirds. Yet just intonation does not quite work either: the comma cannot be dealt with as easily as that. Not to confuse matters, but the comma at issue here is not the Pythagorean comma, which as we saw is the difference between twelve perfect fifths and eight octaves, but rather the syntonic comma, the audible fraction between four perfect fifths and two octaves plus a major third, which may be expressed mathematically by the ratio 81:80. The syntonic comma is also known as the comma of Didymus, after the Greek theorist who discovered it. The difference between the Pythagorean comma and the comma of Didymus is very slight—0.00114326, not to put too fine a point on it. To demonstrate this dilemma, D. P. Walker, the historian whose witty, erudite studies are indispensable reading in the field of Renaissance music theory, suggests an experiment that anyone with access to a violin or cello may perform: play E on the D string in harmony with the open G string, thereby creating a perfect sixth. Now, taking care not to move the finger, play the same E with the open A string, which classical theory would have us believe should produce a perfect fourth. However, if the sixth was truly sweet, then the fourth will be found to be slightly dissonant, and the finger stopping the E will have to be moved forward a considerable amount to tune it. The distance between the two E's is the syntonic comma.

At the same time that the competing claims of the Pythagorean scale and just intonation were being disputed, a third method emerged for resolving this imperfection in the musical intervals. The development of fretted instruments such as the lute and the guitar, and keyboards like the newly invented virginals, demanded a solution if consorts of instruments and voices were ever to be able to perform together. The system that was devised, which has served as the basis of virtually all Western music since the seventeenth century, is called equal temperament, which divides up the syntonic comma among all twelve semitones of the chromatic scale. On a modern piano or guitar, each of the tones that we hear has been slightly jiggered so that the whole gamut of tones will "add up" and thus create very close approximations of all the harmonious intervals. On a modern piano the octaves are perfect, but the fifths, fourths, and thirds and so forth are not, although they are so close to being harmonious that we presumably do not hear the difference. Although a person with normal hearing has the ability to hear the difference between a pure fifth and a tempered fifth, we have become so accustomed to hearing the tempered scale that we have become, in effect, "earwashed": having heard tempered fifths all our lives, they sound right, or right enough, to most of us.

The distinction may seem a bit rarefied today, but in northern Italy in the sixteenth century it was a burning issue that provoked intellectual clashes of a high order. The first great feud in the history of musical science pitted Gioseffo Zarlino, a Venetian musical theorist described by Oliver Strunk as "the most influential personality in the history of musical theory from Aristoxenus to Rameau," against his student Vincenzo Galilei, father of the astronomer. It was a bitter quarrel—vendetta, one is almost tempted to say—centering around the question of temperament. Zarlino passionately defended just intonation in a cappella (unaccompanied) singing, while Galilei as fervently advocated equal temperament, a method of which he was an early pioneer.

Zarlino's arguments were idealistic and arrived at entirely a priori. His reasoning, as summarized by D. P. Walker, ran thus: Nature is superior to art, hence art imitates nature, but nature never imitates art; the human voice is natural, while man-made instruments are not; just intonation is natural, while the system of temperament used by instruments is artificial; therefore, the human voice must use a natural system, that of just intonation, for if it used a tempered scale, that would be an instance of nature imitating art, which is impossible. Logically airtight, yet it demands of the singer a degree of acuity in the matter of pitch much greater than that ordinarily found in nature. To justify Zarlino's logic, singers would have to be capable of distinguishing instantly between the two E's in the little experiment given above, and actually be able to sing slightly different pitches according to the interval they are attempting to harmonize. That is not impossible, but it does mean, in the first place, that there can never be stability in pitch, and secondly, that the voice cannot be tuned with the lute or keyboard—both severe limitations.

Galilei responded with the persuasive argument that all scales are man-made, or, to put it another way, that no musical tone is more natural than another. He was on firm ground there: the human voice is no more free or "natural," musically speaking, than any other untempered instrument, such as a viol, with its unfretted fingerboard, or a valveless horn. The idea that all scales are man-made, and thus on an equal footing, is a provocative one, and although Galilei did not do so, a study of birdsong would have readily confirmed the justness of his assertion. Moreover, the multiplicity and complex variability of non-Western tunings, which were not studied until a later period, also support Galilei's position.

In his *Dialogo della musica antica e della moderna*, which was in large part a vituperative attack on his former master, Galilei asserted that the ear can be trained to hear harmony in tempered, and thus imperfect, intervals. He did agree with Zarlino that a pure, Pythagorean fifth is "more perfect, more sweet than any other form; as I have judged by ear after many, many experiments (since I know of no other better means of achieving certainty in this matter)." Yet he asserted that constant exposure to the new tempered intervals had corrupted human ears, and made mankind actually prefer the "narrow," i.e. tempered, fifth, over the pure interval—a judgment that has been proved correct.

Before we take leave of Gioseffo Zarlino, it ought to be mentioned that while he was among the first of the *musici* to undertake scientific research in the field, he was also a committed proponent of the great theme of cosmic harmony in its pure classical form. In his treatise *Le istitutioni harmoniche*, he sets forth a humanist's précis of *musica mundana*:

> But every reason persuades us to believe that the world is composed with harmony, both because its soul is a harmony

(as Plato believed), and because the heavens are turned round their intelligences with harmony, as may be gathered from their revolutions, which are proportionate to each other in velocity. This harmony is known also from the distances of the celestial spheres, for these distances (as some believe) are related in harmonic proportion, which although not measured by the sense, is measured by the reason.

The Zarlino-Galilei dispute was not quite as stark as I have painted it, for as was the case in most Renaissance vendettas, there was a great deal of cut and thrust, and the two men were blinded by the bitterness of their personal disagreement to the point that, in the case of Galilei, it overpowered his intellectual judgment. At times he adopted positions he did not believe simply because they were opposed to Zarlino. As D. P. Walker writes, "This crookedness, inconsistency, and evasion on the part of Galilei is not only annoying for us, but also unfortunate; for he was an original thinker and a widely experienced musician—he claimed to have collected and entablatured over 14,000 pieces of music, and he had some very interesting things to say; but his line of thought is constantly sidetracked and distorted by his obsessive need to contradict Zarlino."

Galilei did make one unquestionably valuable—and icono-clastic—contribution to musical science in his disproof of Pythagoras's discovery of the inverse proportions in the ratios of the musical intervals. As the reader will recall, according to Iamblichus, Boethius, and all the "official" biographers of the Master, after Pythagoras first made his discovery about the musical intervals in the smithy's yard, he went home and confirmed the numerical congruity of the hammers' weights to the musical intervals by carrying out a number of experiments. He tied weights to gut-strings of equal length, so goes the story, and found in the musical tones thus produced the same simple arithmetical ratios that had been revealed in his original discovery with the weighted hammers: the octave being 2:1, the fifth 3:2, and the fourth 4:3.

That is all wrong, Galilei wrote triumphantly in his treatise *Discorso intorno alle opere de Gioseffo Zarlino*, published in 1589 (which was intended, characteristically, as a definitive refutation of Zarlino's harmonic theory). Galilei revealed that experimental evidence proved that the weights must be in squared, not simple, inverse proportion to the string lengths. In other words, to create the octave-fifth-fourth series, the weight-string relationships would need to be, respectively, 4:1, 9:4, and 16:9, the squares of the numbers of Pythagorean legend. The simple arithmetical ratios work well enough if you are shortening a string, such as a monochord; it is only when you start adding weights to strings of equal lengths that you must square the numbers. Using the sort of rhetorical fig leaf that would become more frequently in evidence as scientists, notably his son, levelled ever more serious challenges to the established wisdom, Galilei said that it was not the Master himself who was at fault but rather his overzealous disciples.

Having got that much right, and in the process endearing himself to us moderns by so clear-cut and heroic a demonstration of the scientific method, Galilei then went on to get himself into trouble. He boldly asserted that the same harmonic intervals can be produced by pipes in a relationship of *cubic* proportions, which gave him a wonderfully tidy mathematical scheme in which musical ratios were determined in two dimensions (string lengths) by a squared series, and in three dimensions (pipes) by a cubed series. Tidy but wrong: for experimental evidence, which had apparently served him so well in rectifying Pythagoras's original error, shows that the pitch of a pipe is the function not of its volume but rather of its length. Galilei himself was behaving like a Greek scientist, telling Nature how she ought to behave rather than consulting her on the matter. If only he had bothered to undertake a bit of empirical investigation, he might have found a more glorious niche in the history of science. As it happened, all the glory would go to his son, Galileo Galilei, the physicist and astronomer, who, though it is not well known, performed some musical experiments of his own, which he published in 1638, when he was a blind and humiliated old man.

Galileo's musical experiments are to be found in a dialogue entitled *Discorsi e dimostrazioni matematiche*. In the middle of a long discourse devoted to pendulums, Galileo repeated his father's discovery about the squared relationship between the harmonic musical intervals and the ratios of weighted gut-strings, which would tend to suggest that Vincenzo's work had failed to make an impact. One of the speakers in the *discorso*, Sagredo, challenges the conventional wisdom that the standard ratios, 2:1, 3:2, and so forth, correctly express the *forme naturali* of the musical consonances. Sagredo, who is Galileo's mouthpiece, asserts that there are three ways of altering the musical pitch of a string: by shortening it (the monochord method), by making it taut with weights, and by altering its thickness. He then asserts exactly what his father had already proved, that the ratios obtained with a monochord are in simple inverse proportion, while the ratios obtained by the second and third methods are in inverse squared ratios. (Apparently, Galileo obtained his results for the third method by comparing the area of a section of the string; in other words, a string with a cross section measuring four square millimeters, say, will have a pitch one octave lower than a string having a cross section of one square millimeter.)

In order to verify these assertions, Galileo carried out some experiments that, it would seem, slumbered in obscurity in the depths of his *Opere* for three hundred and fifty years, until D. P. Walker exhumed them and attempted to reproduce their results. The first experiment calls for a musical note to be produced by rubbing the rim of a glass of water. Galileo promises that you will see "the waves in the water of exactly equal form";

then, if the note suddenly jumps up an octave, "there will appear other, smaller waves, which with infinite precision cut in half the first ones." Although the dialogists say that they have successfully performed the procedure a number of times, Walker discredits the experiment. In the first place, the frequency of sound waves is too rapid to be observed by the eye; one can no more measure the waves produced on a glass of water in this way than one can count the vibrations of a plucked string. And he rejects it in the second place "because I have not yet succeeded in making a sounding glass jump an octave." Nor have I; the reader is invited to have a go at it.

The second experiment, Galileo writes, he hit upon by chance while scraping a brass plate with a chisel in order to remove spots from it. He discovered that when he struck the chisel in just the right way, it made a high-pitched musical sound as it scraped across the plate; afterward he found a series of minute striations in the plate, parallel and equidistant—to him obviously equivalent to the waves in the water-glass experiment. By practice, he says, he was able to produce notes with the chisel and brass plate that were exactly a fifth apart, a relationship he verified by playing the notes on his harpsichord (though he does not tell us whether his harpsichord was tempered or not). Lo and behold, when he compared the number of striations made by the two marks, he found forty-five lines in one and thirty in the other, "which truly is the form attributed to the fifth."

Even setting aside the fact that the experiment is not readily reproducible, to put it in the mildest terms, the principal problem is that Galileo's experiment does not take into account the time element, which is crucial. The number of striations, meant to be a graphic representation of the sonic vibrations that produce musical tones, is meaningless unless the scrape marks are produced in exactly the same space of time. Walker charitably concludes that this "was one of Galilei's 'thought-experiments,' about which he had not thought quite enough, though he tells the story with wonderfully convincing realism. I was greatly relieved to notice this mistake, since otherwise I should have had to waste a lot of time ineffectively scraping brass plates with chisels."

One final point about Galileo's musical science: in order to explain *why* consonant musical tones sound harmonious, he supported a theory, propounded since at least 1585, that harmony is produced when the sound waves of two or more notes coincide. Thus an octave (with the ratio of 2:1) is the most harmonious of all the intervals, for the sound waves produced coincide with every other "wave"; next is the fifth (3:2), in which the two notes coincide every third pulse of the higher note; next the major third (5:4), coinciding every fifth pulse. He "proved" this theory, or rather illustrated it graphically, with pendulums set at speeds to correspond with the frequency of the postulated sound waves. The result, said Galileo, was "a beautiful entwining," in which the eye could "take pleasure in seeing the same games that the ear hears."

Going a step further and adding an aesthetic element, Galileo asserted that the theory also explains why it is that less perfect intervals produce less pleasurable consonances. The octave, he said, is "sdolcinata troppo e senza brio," too sweet and lacking in verve, because the repetitive two-to-one ratio produces a boring alternation of sound-wave coincidences, a sort of pitch equivalent to the rhythm of an oompah band. A fifth, on the other hand, produces a more sophisticated rhythm, which "makes such a tickling and stimulation of the cartilage of the eardrum that, tempering the sweetness with a dash of sharpness, it seems delightfully to kiss and bite at the same time."

It gives no pleasure to dash cold water on so charming a theory, but the whole notion of coincident waves is groundless. The question of the ratios of the harmonious musical intervals was to be taken up again at the end of the seventeenth century, by none less than Newton himself. We shall examine what he has to say on the subject in due course, but let us now turn our attention from these first scientific attempts to elucidate the ultimate truths of musical harmony to trace how the great theme began to enter the mainstream of the culture.

SIX

The Music of the Spheres and the Birth of the Opera

It is surprising that Vincenzo Galilei does not occupy a more prominent position in the history of music. If he is mentioned at all, he is usually dismissed as a contentious buffoon completely overshadowed by his brilliant son. Yet in his work we may find clearly outlined not only the new scientific direction of music theory, as we saw in the last chapter, but also the incipient rise of profane musical composition. It had always been universally accepted that the *musici*, the theorists, were the true musicians, and that the actual music makers, particularly the composers and performers of profane music, were scarcely more to be taken seriously than were gymnasts or clowns. All that was about to change forever.

Vincenzo Galilei's stated purpose in the *Dialogo della musica antica e della moderna* was to prove the superiority of the musical practices of the ancients over the corrupt and decadent music of his own time. Once again, we have that paradox so typical of the Italian Renaissance: genuinely new thinking disguised as a re-discovery of the classical era. In the introduction to his dialogue, Galilei takes pains to identify himself as a thoroughly modern man by a reference to the New World: "After Italy had for a long period suffered great barbarian invasions, the light of every science was extinguished, and as if all men had been overcome by a heavy lethargy of ignorance, they lived without any desire for learning and took as little notice of music as of the western Indies."

Galilei offers as evidence of the superiority of ancient music the extensive body of classical literature asserting the miraculous powers of music to move the spirit, heal the body, and otherwise influence sublunary affairs. He took all these accounts very much at face value, to the point that his principal proof of the inferiority of modern music is the fact that the polyphony of his day was incapable of working the marvels that Orpheus, Pythagoras, and the other great figures of antiquity were able to accomplish with their music making. The concept of a lost golden age played an important role in the humanism of the Renaissance. It was clearly present in the dreamy arcadias painted by Botticelli and Giorgione, and in Raphael's visions of the academy of ancient Greece, as well as in the idealistic verses of Dante and Petrarch. As we shall see, it was also the essential motivation behind the magical hermetism of Marsilio Ficino and Pico della Mirandola, the Florentine humanists whose musical theories were the connecting link between antiquity and modernity. All these were manifestations of a yearning for a lost epoch of excellence, virtue, and mystical wisdom. "The ancient music, I say," wrote Galilei, "was lost, along with all the liberal arts and sciences, and its light has so dimmed that many consider its wonderful excellence a dream and a fable." In the *Dialogo* his ultimate purpose is the restoration of the music of the golden age.

The particular objects of his wrathful contempt were the practitioners of polyphony; indeed, polyphony was essentially how he defined the *musica moderna* of the title. Strongly influenced by a humanist named Girolamo Mei, a learned linguist and student of Greek drama with whom he corresponded, Galilei believed that antique music was monodic, that is, consisting of a single singing voice, simply accompanied by lyre or other instruments, and thus the antithesis of polyphony. The emphasis was on the words, which were able to speak directly to the soul, unencumbered by irrelevant musical artifices. Mei and Galilei were basically correct in their assumptions about Greek monody, although the former seems to have supplied his colleague with some thirteenth-century Byzantine musical manuscripts in the belief that they were of great antiquity. Galilei reserved particular scorn for those of his contemporaries who sought to beguile their listeners with the sensual delights of polyphony (the term was used to describe instrumental consorts with several parts as well as purely vocal compositions). He expounded this view in two very long and sarcastic sentences in the *Dialogo*:

> If the object of the modern practical musicians is, as they say, to delight the sense of hearing with the variety of consonances, and if this property of tickling (for it cannot with truth be called delight in any other sense) resides in a simple piece of hollow wood over which are stretched four, six, or more strings of the gut of a dumb beast or of some other material, disposed according to the nature of the harmonic numbers, or in a given number of natural reeds or of artificial ones made of wood, metal, or some other material, divided by proportioned and suitable measures, with a little air blowing inside of them while they are touched or struck by the clumsy and untutored hand of some base idiot or other, then let this object of delighting with the variety of their harmonies be abandoned to these instruments, for being without sense, movement, intellect, speech, discourse, reason, or soul, they are capable of nothing else. But let men, who have been endowed by nature with all these excellent and noble parts, endeavor to use them not merely to delight, but as imitators of the good ancients, to improve at the same time, for they have the capacity to do this and in doing otherwise they are acting contrary to nature, which is the handmaiden of God.

As bizarre as it will seem to us today, all of this earnest exhortation to simplicity and spiritual self-improvement led directly to the birth of the opera, the most extravagant and voluptuous form of musical entertainment ever devised. The principal speaker in Galilei's *Dialogo*, and its dedicatee, was a wealthy Florentine patron, "the very illustrious Signor Giovanni Bardi de' Conti di Vernio, who having long studied music . . . has greatly ennobled it and made it worthy of esteem, having by his example incited the nobles to the same study, many of whom are accustomed to go to his house and pass the time there in cultivated leisure with delightful songs and laudable discussions." Bardi's salon, which was known as the Camerata, commissioned and produced the first operas. The argument over which of the works produced by members of Bardi's Camerata ought properly to be called the first opera will never be settled, but there is little doubt as to the most important predecessor of the new medium. The first significant work in the *nuova maniera di canto*, as it came to be called (*nuova* again, that favorite adjective of the Renaissance), was a suite of musical interludes, known as the *Pellegrina* intermedi, which were composed for the wedding of Ferdinand de' Medici and the princess Christine of Lorraine, in 1589.

Ferdinand had become the Grand Duke of Tuscany in 1587 after the death of his brother, the Grand Duke Francesco de' Medici; rumor had it that Ferdinand poisoned his brother for the ducal throne. In any case, Ferdinand gave up a cardinal's mitre in order to marry the French princess, a granddaughter of Caterina de' Medici, for he hoped thereby to consolidate power in his somewhat shaky dukedom. A sumptuous and impressive wedding was vitally important to him, so he engaged Giovanni Bardi to supervise the festivities. The high point of the celebration were the intermedi, the most spectacular event in the history of European music up to that point.

It is a bit misleading to call the Florentine intermedi the forerunners of the opera, as though they are nothing more than embryonic stages, marking the inexorable passage upward to operahood, for they had a formal character and artistic interest all their own. Nonetheless, they did have all the essential features of the emergent *nuova maniera*, especially the mixture of styles—that is, recitative, the Camerata's attempt to recreate the music-accompanied dialogue of classical drama, and aria, melody sung by a single voice—that would dominate opera until Wagner's "endless melody" 250 years later.

The intermedi were loosely connected musical tableaux, very much like the numbers in a modern musical revue, performed between the scenes of a stage play. As a rule, there was no particular connection between the intermedi and the play. On the occasion of Ferdinand and Christine's wedding, the play was Girolamo Bargagli's comedy *La Pellegrina*, a romantic farce involving the preposterous ironies and mistaken identities typical of Mannerist comedy. The theme of the intermedi was "The Power of Music," a suitably serious and classical subject. To illustrate his theme Bardi employed the services of the best composers and poets in Tuscany. Among the former were Peri, Caccini, Cristoforo Malvezzi, and Emilio de' Cavalieri, who in 1600 composed an allegorical work entitled *La rappresentazione di anima e di corpo*, which is sometimes called the first oratorio. The chief poet of the entertainment was Ottavio Rinuccini, who would later write the books for the first operas. The sets were the work of Bernardo Buontalenti, a leading artist of the day, who designed fabulous stage machines that required eight months—and a budget Franco Zeffirelli might envy—to construct. Buontalenti's stage magic included a Mount Olympus that rose from beneath the stage, a flying car drawn by winged dragons breathing fire and smoke, a python that whistled and wiggled on the ground as it vomited sparks, and a ship with forty sailors that seemed to sail upon the sea. The spectacle, which was performed on May 2, 1589, in the Grand Royal Salon of the palace (now part of the Uffizi Gallery), was an enormous success. Ferdinand was so delighted with the production that he instantly demanded that it be repeated and, fortunately for posterity, ordered that the music be published. The *Pellegrina* intermedi are the only examples of the genre that survive in a complete form.

The intermedi are six in number. The first and last of them are the most succinct summation of the great theme of the harmony of the spheres to be found in music literature. The first intermedio, entitled "L'armonia delle sfere," is a straight-forward rendering of Plato's Myth of Er, which would have been familiar to everyone in Bardi's audience, while the sixth one, also inspired by Plato, represents Jove's gift to mortals of rhythm and harmony. The second number depicts the singing contest between the nine Muses and the Pierides, the daughters of Pierus; the Muses win, and the Pierides are turned into magpies. The third piece, Apollo slaying the python at Delphi, seems to have been devised mainly in order to accommodate Buontalenti's dragon, and is a bit of a stretch, in terms of fitting the theme of the Power of Music: after the god has killed the dragon, the Delphians rejoice with music and dance. In the fourth intermedio, a sorceress summons up demonic spirits with her song and foretells the return of the Golden Age. The fifth number, borrowing a theme straight out of Pythagorean legend, tells the story of the poet Arion, who was rescued from murderous sailors by a music-loving dolphin that was attracted by his singing. A detailed description of the intermedi comes to us from Bastiano de' Rossi, a participant in the festivities who published an account of them a few months after the performance—a good indication of how successful the specta-cle was. Here is how Rossi describes the first number, "L'ar-monia delle sfere":

> This intermedio represented the celestial Sirens, led by Harmony, as mentioned by Plato in his *Republic*. In addition to the ones mentioned by him, we added two, in accordance with modern thinking; which is to say, the ninth and tenth

spheres. When the curtain fell* you could see immediately a cloud appearing in the sky, and almost in front of the stage was a small temple of the Doric order, in rustic stone. A woman was descending on the cloud very slowly. She was playing the lute and singing the following madrigal:

> From the highest sphere,
> friendly escort of the celestial Sirens,
> come I, Harmony, to you, O mortals:
> for the winged messenger has brought to Heaven
> a report of the greatest importance,
> that the Sun never saw a couple noble as you,
> O new Minerva and strong Hercules.

Then the backdrop fell away, revealing the Sirens, disposed on a number of clouds, singing and playing harps, viols, lutes, and *chitarroni*, a sort of bass lute. Rossi continues, "In the sky and on earth there came to be heard a melody so sweet, such as perhaps had never been heard before, as it were from Paradise":

> We, by whose singing the celestial spheres
> are sweetly made to turn round,
> have on this happy day
> taken leave of Paradise
> to sing the greater wonders
> of a beautiful spirit and a comely face.

Then three more clouds descended to the scene. Enthroned on the largest one was Necessity, clasping the cosmic spindle between her knees, in a literal representation of Plato's Myth of Er. Below her were the three Fates. The celestial orchestra was further swelled by the planets and a group of heroes, which included a boy who sang a short solo. (The boy was added at the insistence of Duke Ferdinand, who was especially fond of his

---

* In the sixteenth century, it was customary for the stage curtain to fall, rather than to be raised, at the commencement of a dramatic performance.

voice.) The instrumentation of the assembled heavenly host was expanded to include *cornetti*, trombones, flutes, a mandolin, and a psaltery.

Cavalieri, the orchestra leader, kept notebooks which have survived, with details of the performance. From him we know that the instruments of the stage musicians in the first and sixth intermedi were painted with golden sunbeams; while in the infernal fourth intermedio the viols and trombones played by the sorceress's demons were covered with green silk, to match their costumes.

The sixth intermedio, the full title of which is "Il dono di Giove ai mortali di Ritmo e Armonia" (Jove's gift of Rhythm and Harmony to mortals), is, scenically, a virtual reprise of the first tableau, which gives the *Pellegrina* intermedi a classical symmetry. The curtain falls upon a dawn sky. Sunlight streams from behind Mount Olympus, where Jove sits enthroned. Floating on clouds around him are the Graces, the Muses (triumphant after their victory over the Pierides in the second intermedio), four cupids, Apollo and Bacchus, and Harmony and Rhythm. This divine chorus sings its praises to Jove for his gift to humanity of music and dance, and summons mortals on earth to come receive the blessing of heaven:

> *O how brightly do the clouds shine*
> *in the air, and with what beautiful colors!*
> *Hasten hither, shepherds,*
> *and you, charming, happy,*
> *lovely nymphs, hasten hither*
> *to hear the sweet sound*
> *of celestial harmony.*

The donation of music and dance is made, and then mortals and gods join together in a celebratory *ballo*, a stately dance that was inserted at Ferdinand's request. The mortals ask, "Will the Age of Gold return?" and the reply from heaven is, "The

Golden Age will return, and with it royal ways will be ever more brilliantly enlightened." In this Golden Age, the oak trees will drip with honey and the rivers flow with milk. As befits a royal wedding celebration, the credit for all this earthly joy is attributed to the royal pair; nymphs and shepherds sing the glory of the Arno to the skies, and heaven and earth together sing the praises of Ferdinand and Christine.

The most surprising thing about the intermedi, after one has read Galilei's avowals of the Camerata's intention to revive classical monody, is the extensive use of polyphony. The music for the *Pellegrina* intermedi was, if anything, even more elaborate than what had preceded it. The explanation, of course, is that the passionate interest in ancient music on the part of Galilei and the other members of the Camerata was more idealistic than practical. Few of the principles so earnestly articulated in their treatises were put into practice in their music; and there is certainly no formal resemblance at all between the compositions of the Camerata and ancient Greek music. It is a truism to say that art can only hold the mirror up to its own time; but, like most truisms, it is surely true.

Beginning around 1600, the Camerata produced a number of mythological music dramas based upon the principles established by the *Pellegrina* intermedi, especially the mixture of recitative and accompanied melody. Peri's *Dafne*, setting a poem by Rinuccini, was the first work to conform to the modern definition of opera, but the music is now lost. After that things get a bit more complicated: in 1600, Caccini published the music for his opera *Euridice*, also composed to a libretto by Rinuccini, but it was not performed until 1602, at the Pitti Palace. In 1600, meanwhile, Peri produced his own version of *Euridice*, using the same text and some of Caccini's music. Thus the question of what was the first opera takes three answers: the first performed, the first published, and the oldest surviving operas are, respectively, *Dafne*, Caccini's *Euridice*, and Peri's *Eu-*

ridice. It is a largely academic question: not until we reach Monteverdi's *Orfeo*, first produced in Mantua in 1607, do we encounter an opera with the compelling dramatic interest that we now associate with the medium.

The ballet in western Europe evolved at precisely the same moment as the opera, and the great theme found prominent expression in it as well. The dance spectacle usually identified as the first ballet in the modern sense was the *Ballet comique de la Reine*, which, although produced in France, was another entertainment performed in celebration of a wedding at a Medici court, in this case that in Paris of Caterina de' Medici, the wife of one king of France and the mother of three (as well as the grandmother of Christine, Duke Ferdinand's bride). Like all of that ilk, Caterina de' Medici had a pronounced proclivity for pomp, luxury, and the dernier cri in matters of art and taste. When the sister-in-law of her son, King Henry III, was affianced to the Duc de Joyeuse, a series of lavish spectacles was performed, including allegorical tournaments, musical concerts, masquerades, a water fete, a "horse ballet," and a magnificent fireworks display. The climax of the festivities was the *ballet de cour*, which was devised by Balthasar de Beaujoyeulx, a stage director and experienced musician from Savoy; one contemporary critic called him the greatest fiddler in Christendom.

On the night of the ballet, September 18, 1581, the Salle de Bourbon was mobbed by thousands of Parisians who had heard that something unprecedented was about to take place. Although admission was supposed to be granted only to "persons of marque," the salon was thronged with an overflow crowd of thousands of people. The music was provided by Lambert de Beaulieu, a composer associated with the newly founded Academy of the poet Jean-Antoine de Baïf, one of the Pléiade, who, like the humanists at the Medici court in Florence, was deeply engaged in the business of trying to reconstruct classical Greek drama and music. Thus, as Frances Yates writes in her detailed study of the *Ballet comique*, "The music and dancing of this performance were of the 'ancient' kind, and related to the harmony of heaven."

The settings for the ballet were quite as extravagant as those for the Florentine intermedi. On one side of the room was an oak grove in which was seated the god Pan; behind him was a grotto that concealed a musical consort. The grove was lighted by silver lamps in the shape of ships, which hung from the boughs of the trees. At the opposite end of the room was a wooden vault, laminated with gold and very brilliantly illuminated, and surrounded by billowing clouds. Called the *voûte dorée*, it too was filled with singers and players. Concerning the golden vault, Beaujoyeulx wrote that those in the audience who were "better instructed in the Platonic discipline esteemed it to be the true harmony of heaven, by which all things that exist are conserved and maintained."

The *voûte dorée* was less explicit a representation of the great theme than that presented by the Florentine intermedio "L'armonia delle sfere." Likewise, the action of the ballet, a series of somewhat garbled stories from Homer, was on the surface far removed from the principles of cosmic harmony, although Beaujoyeulx did on several occasions allude to the power of music to tame the passions, and suggested that harmony in the soul is patterned after the harmony of the universe. It was in the finale, the Grand Ballet, that the *Ballet comique de la Reine* clearly and graphically expressed the principles of the theme of cosmic harmony. It was also that portion of the entertainment that historians of the dance have regarded as the first modern ballet, for the Homeric tableaux that preceded it were musical dramas in the established tradition of the court masque. The Grand Ballet was a dance of triumph, enacted by naiads and dryads celebrating the defeat of the evil enchantress Circe. Frances Yates describes the Grand Ballet and its place in the development of the great theme in her book *The French Academies of the Sixteenth Century*:

The Grand Ballet was composed of forty passages, all geometrical figures, all most exactly and accurately described—now square, now round, now a triangle, now some mixed figure. As each figure was marked by the twelve naiads, the four dryads broke it and formed another. Halfway through there was a grand chain formed of four different kinds of interlacings "and so dexterously did each dancer keep her place and mark the cadence that the beholders thought that Archimedes himself had not a better understanding of geometrical proportions than these princesses and ladies showed in the dance." The Pythagorean-Platonic core of the Academy—that all things are related to number, both in the outer world of nature and in the inner world of man's soul—perhaps found in the marvelous accuracy of this measured dancing one of its most perfect artistic expressions.

One could argue that by the time of the *Ballet comique de la Reine*, in 1581, and the *Pellegrina* intermedi eight years later, what we have been calling the great theme of cosmic harmony had become exactly that, a theme and nothing more, and such toadying court entertainments as these were little more than watered-down repetitions of the Pythagorean-Platonic beliefs, intellectually and spiritually very far from the profound sublimity of the originals. It would be difficult to confute the argument entirely. At the same moment that the court ladies of Paris were prettily forming geometric figures in the Salle de Bourbon, and the Camerata were flattering the Grand Duke of Tuscany that he was ushering in a new Golden Age, modern science was being born. However, it is only with the hindsight of centuries that we are able confidently, perhaps smugly, to say that at X point in time began the intellectual movement that would ultimately disprove the premises of the Pythagorean-Platonic view of the harmonious cosmos. Likewise, it ought to be borne in mind that the Camerata did not convene with the express intention of creating a new and prestigious musical medium; nor did Balthasar de Beaujoyeulx call together the artistic elite of Paris

and say, "Let's invent the ballet." They were simply expressing
artistically their own beliefs, necessarily the beliefs of their age,
as ingeniously and entertainingly as they could.

It is undoubtedly true that the *Pellegrina* intermedi and the
Ballet comique de la Reine mark the beginning point of the widespread propagation of the great theme—and thus, arguably,
the point at which it ceases to have intellectual bite. A parallel
closer to our own time is the influence of Christianity in the
latter nineteenth century. From a contemporary perspective it is
obvious that by that time Christianity had ceased to be the vital
leading edge in Western thought: the best minds were engaged
in proposing new ideas that would lead to doubt and a decline
in orthodox faith. Any modern intellectual history would cite as
the great and influential thinkers of that age Darwin and Freud,
Marx and Nietzsche, not the sermonizers of established Christianity. The images from that era that have gained prominence
are the sensual materialism of the French Impressionists and
the luxurious fantasies of decorative artists such as Whistler
and Toulouse-Lautrec, not the treacly religious scenes of a
hugely popular artist like William Holman Hunt. The modern
critical judgment is that the greatest French poets of that period
are the violently anti-establishment Baudelaire and Rimbaud;
looking across the Channel, we now teach undergraduates
about Matthew Arnold's perception of the "melancholy, long,
withdrawing roar" of the Sea of Faith, while we ignore the vast
catalogue of insipid inspirational verse of the period.

Let us accept all that as obvious. Yet who would be foolish
enough to deny that Christianity was the single most important
social force in Europe in the late nineteenth century? Its beliefs
were accepted unquestioningly by most educated people; its
power supported every social institution; it was the driving force
behind the imperialism that was transforming the rest of the
world, with repercussions that are still being intensely felt today.
Most of the intellectual revolutionaries mentioned in the previous paragraph were unknown to the great majority of people

living at that time, and those which were known were, for the most part, despised and ridiculed. I do not want to overemphasize this parallel, for it is not apt in every respect; however, it would not be an exaggeration to say that the Pythagorean-Platonic conception of the musical universe was just as important to the Renaissance and to the ages that followed, almost to the end of the eighteenth century, as Christianity was to the Victorian age.

Nonetheless, it is quite true that as the new science began to emerge, treatments of the great theme shift more and more into the realm of the imagination. In the late Renaissance and the period that followed, the concept of the harmonious universe is apparent in art and literature with the same ubiquity with which it was found in the scientific and religious treatises of the classical era. Several scholarly books have been written tracing the images of cosmic harmony in art and literature, and there is no need to duplicate their efforts here. Yet it is a commonplace among students of Elizabethan literature, for example, that every educated person of the period accepted as an article of faith the premise that the heavens were composed of crystal spheres, which made a harmonious music as they were moved round by angels. Lorenzo's famous summation of the great theme in *The Merchant of Venice* may stand for a multitude of such instances:

> How sweet the moonlight sleeps upon the bank!
> Here we will sit, and let the sounds of music
> Creep in our ears: soft stillness and the night
> Become the touches of sweet harmony.
> Sit, Jessica: look, how the floor of heaven
> Is thick inlaid with patines of bright gold:
> There's not the smallest orb which thou behold'st
> But in his motion like an angel sings,
> Still quiring to the young-eyed cherubins.
> Such harmony is in immortal souls;
> But, whilst this muddy vesture of decay
> Doth grossly close it in, we cannot hear it.

In the Elizabethan era, belief in *musica humana*, no less than in *musica mundana*, was universally accepted, and came to be codified in something called the Great Chain of Being. The Great Chain derived directly from the concept of cosmic order that underlay the elaborate structure of Pythagorean-Platonic thought. As the modern scholar of Elizabethan literature E. M. W. Tillyard points out, the Great Chain of Being was "one of those accepted commonplaces, more often hinted at or taken for granted than set forth." One concise summation, from the fifteenth-century English jurist Sir John Fortescue, may stand for the rest: "In this order angel is set over angel, rank upon rank in the kingdom of heaven; man is set over man, beast over beast, bird over bird, and fish over fish, on the earth, in the air, and in the sea: so that there is no worm that crawls upon the ground, no bird that flies on high, no fish that swims in the depths, which the chain of this order does not bind in the most harmonious concord." The last phrase clarifies and makes explicit the musical analogy implied by the Great Chain of Being.

We may quote another, particularly lovely expression of the great theme from Milton, a century later, if only to show that long after what modern historians have labelled the Copernican Revolution an intellectual heavyweight still had implicit faith in the ancient theme of celestial harmony—a chain of belief unbroken from the first, defining moments of Western civilization. In his lyric poem "Arcades," Milton's Genius of the Wood explains to an assembly of nymphs and shepherds how he listens to the music of the spheres at the end of the day:

> But else in deep of night, when drowsiness
> Hath locked up mortal sense, then listen I
> To the celestial sirens' harmony,
> That sit upon the nine infolded spheres
> And sing to those that hold the vital shears
> And turn the adamantine spindle round,
> On which the fate of gods and men is wound.

> *Such sweet compulsion doth in music lie,*
> *To lull the daughters of Necessity,*
> *And keep unsteady Nature to her law,*
> *And the low world in measured motion draw*
> *After the heavenly tune, which none can hear*
> *Of human mold with gross unpurgéd ear.*

Milton, no less than Pythagoras himself, made it his business to unpurge human ears, albeit in a vastly more complicated world. The great theme still had another century to run before it finally bowed to the combined forces of science triumphant and the rise of Romanticism. Yet the intellectual forces were already afoot that would drive it underground.

SEVEN

The Hermetic Tradition

Since the time of its founding, Pythagoreanism was at its core a mystical institution. Closer to a religion than a school of philosophy in the modern sense, the Brotherhood depended upon a process of initiation and revelation: the oracular wisdom of Pythagoras was transmitted only to those fortunate few who were admitted to the cult as adepts. A further quality of the Pythagorean tradition is that it was all-embracing. For its adherents, the Master's teaching was not merely a matter of spiritual enlightenment; it also contained in its codified wisdom everything one needed to know about science, art, law—the whole gamut of earthly life—though it is quintessentially un-Pythagorean to divide the world up that way. The idea that distinctions ought, or even could, be made between religious and temporal law, or between science and art is a modern, or at least a post-classical notion that is at odds with the Pythagorean vision of cosmic harmony and unity. As Plato had exhaustively demonstrated in the *Timaeus*, music and mathematics could be used to explain every terrestrial phenomenon; they encompassed every possible dimension of thought.

The tradition of the infallible and all-embracing doctrine

was widely broadcasted by the medieval scholastics and encyclopaedists. Yet the devotees of the Pythagorean cult despised such vulgar propagations, not because of their crude oversimplifications of sublime principles, or even their outright errors, but for the very fact that they exposed the words of the Master to the uninitiated in the first place. The codified wisdom of the cult, the so-called Golden Verses of Pythagoras, were holy and not meant to be exposed to the profane gaze of the uninitiated. Secret mystical societies on the Pythagorean model were common throughout the classical world. Even as the principle of the great theme of cosmic harmony was becoming accepted as an essential intellectual assumption throughout the culture at large, such groups flourished under the rose. During the Renaissance, while investigative music theory of a scientific nature was being explored by the likes of Zarlino and Vincenzo Galilei and his son, the ancient mystical tradition lived on in Neoplatonic academies that were the period's equivalents of the Pythagorean-style mystery cults. In order to trace the survival of the esoteric tradition, we must revert to the early days of the Renaissance, about a hundred years before the birth of the opera, to the time when patrons such as Cosimo de' Medici began the rediscovery of classical learning that we call humanism.

While the esoteric tradition has never been as prevalent or as powerful as its adherents in different ages (including our own) have claimed, it is nonetheless probably true to say that it has existed in one form or another since the time of Pythagoras himself. With the waning of orthodox paganism and the decline of the Greek dialecticians, cults and mysteries sprang up throughout the Mediterranean. In the second century A.D., mystery religions such as the cult of Mithras enjoyed a flowering; mystical Christianity in all its multifarious and heretical manifestations, gnostic and otherwise, began to take hold; and at the same time there was a widespread revival of the cult of Pythagoreanism. We know that these clandestine groups continued to prosper throughout the Middle Ages, for Augustine and the other church fathers wrote about them often, alternately railing against their impiousness and finding in their enigmatic texts divinely inspired prophecies of the coming of Jesus. First among these pagan seers and prophets, even more venerable than Pythagoras himself, was the legendary Hermes Trismegistus, Hermes the Thrice-great. The two big books ascribed to Hermes, the *Asclepius* and the *Corpus Hermeticum*, amounted to an occult encyclopaedia that dealt systematically with astrology, the secret powers of plants and stones, talismans to summon forth airy spirits and demons of the underworld (and charms to ward them off), as well as philosophical literature of a distinctly Pythagorean cast.

Hermes Trismegistus was supposed to have been an inspired Egyptian seer who lived and wrote at the very dawn of antiquity: he was indeed the inventor of writing with hieroglyphs, and thus the father of human civilization. While the true authorship of the Hermetic texts (it is from this Hermes that the word originated) will never be known, the important point is that until the seventeenth century they were universally believed to date from the earliest era of human history. In *The City of God* (*De civitate Dei*) Augustine affirms this notion unequivocally: "For as for morality, it stirred not in Egypt until Trismegistus's time, who was indeed long before the sages and philosophers of Greece." Other ecclesiastic authors contributed their bit to the Hermetic legends. That same Clement of Alexandria who extolled the quasi-magical powers of the New Song of Jesus also wrote a memorable description of the processions of the ancient Egyptians, in which the priests sang magic hymns by Hermes and carried copies of his works as talismans. Clement stated that Hermes Trismegistus was the author of forty-two books, thirty-six of which comprised the essence of sacred Egyptian philosophy. Augustine, Clement, and other early church fathers invoked Hermes's proto-

Christian moralism to persuade pagans to accept Christianity; during the Renaissance the same relationship was asserted to justify the veneration of ancient mysticism.

The reason Hermes Trismegistus concerns us is the enormous importance attached to the rediscovery of the *Corpus Hermeticum* by the Renaissance humanists, to whom it was a major philosophical justification for the revival of the great theme of cosmic harmony, and who perused its magical pages for instruction in how to use music to divert cosmic energy toward earthly ends. As we have seen, one of the deepest currents in Renaissance thought was the desire to find prototypes for intellectual and artistic activity in the golden age of the classical era; and no seer of antiquity was held in higher esteem in the fifteenth century than was Hermes Trismegistus. "Rediscovery" is perhaps not quite the right word, for the devotees of the Hermetic cults had venerated these works, presumably without interruption, down through the centuries. Yet secrecy was, after all, one of the essential attributes of these societies, and for the great majority of people throughout the Christian era until the Renaissance, the tantalizing allusions of the church fathers were as close as they ever got to the mysterious Thrice-great Hermes.

Then, around 1460, Cosimo de' Medici got his hands on an almost complete manuscript of the *Corpus Hermeticum* in Greek. He handed it over to Marsilio Ficino, the brilliant young classicist and philosopher who was foremost among the humanists at Cosimo's Platonic academy. Ficino's life work was the attempt to reconcile Christianity with the pagan classics, a great many of which he translated for the first time out of the Greek. Ficino thus occupies an essential position in the great chain of musical philosophers, for it was he who translated the key works of Plato and the Neoplatonists. When the *Corpus Hermeticum* was brought to Florence from Macedonia, where a monk in Cosimo's employ had discovered it, Cosimo ordered Ficino to set aside Plato's dialogues, which he was working on, in order to translate the new manuscript straight away. Cosimo was then in his seventies, and it seems that he wanted to read the oracular works of the great Egyptian seer as a guide to the afterlife.

The discovery of the Hermetic texts created a sensation in Florentine intellectual circles. Ficino placed Hermes even earlier than had Augustine, making the Egyptian seer roughly contemporaneous with Moses. Thus the Hermetic writings were, with the Pentateuch, which was then ascribed without question to Moses himself, the most ancient, and therefore the holiest, writings of the human race. The question of dating was vital, because Ficino identified many similarities between Genesis and passages of the *Corpus Hermeticum*.

Even more critical was the relationship between Hermes Trismegistus and Plato, because of the close philosophical similarities between the two. By applying a modern critical eye to the texts, it is clear that the Hermetic writings were based upon late, corrupt Neoplatonic texts by writers who were also, obviously, familiar with Genesis. However, Ficino, believing implicitly in the antiquity of the *Corpus Hermeticum*, constructed a philosophical genealogy of what he called the *prisca theologia*, or ancient theology, at the center of which was the revelation of cosmic harmony, made manifest to the ancients but garbled and corrupted through the succeeding ages. In the dedication of his translation of the *Corpus* (to Cosimo de' Medici, of course), Ficino carefully delineated that genealogy: "[Hermes Trismegistus] is called the first author of theology: he was succeeded by Orpheus, who came second amongst the ancient theologians: Aglaophemus, who had been initiated into the sacred teaching of Orpheus, was succeeded in theology by Pythagoras, whose disciple was Philolaus, the teacher of our Divine Plato. Hence there is one ancient theology . . . taking its origin in Mercurius [Hermes] and culminating in the Divine Plato." Later, in his commentaries on Plotinus, Ficino amended the genealogy to include Zoroaster as co-founder with Hermes of the *prisca theologia*.

As much as Ficino and the other Renaissance humanists tried to make pagan mystery religions respectable by casting them as the precursors of Christianity, and however noble and benign some passages of the Hermetical writings might have been, they were still purely pagan and devoted to the practice of magic. It was common knowledge throughout the Middle Ages that the *Asclepius* was a source of charms and talismans that were demonic in nature. Among other things, it instructs its readers that man "has familiarity with the race of demons," which acquaintanceship permits him to make his own gods. The statues of the deities in the temples of ancient Egypt, it seems, were endowed with *sensus* and *spiritus*, which enabled them to speak and move. They could foretell the future and inflict disease and other hardships on humankind, and cure them as well. The priests who made these marvelous images could not actually create souls themselves, says the author of the *Asclepius*, so "after having evoked the souls of demons or angels, they introduced these into their idols by holy and divine rites, so that the idols had the power of doing good and evil." These man-made gods were compounded from herbs, stones, and essences that would attract the proper celestial influences. In order to petition the intercession of these idols, the *Asclepius* instructs its readers to offer sacrifices and to make music, hymns of supplication in harmony with the music of the cosmos.

Ficino himself was an ardent believer in sympathetic magic, and he wrote several treatises on the subject. He was especially drawn to the musical aspects of magic. In his *De vita coelitus comparanda* (which as D. P. Walker points out may be translated both as "On life led in a heavenly manner" and "On obtaining life from the heavens"—and, given Ficino's penchant for wordplay, it probably means both), he offers practical advice on how best to invoke celestial influences. Basing his magic theory on commonly accepted medical assumptions of his day, Ficino states that the spirit is the part of man most directly influenced by the heavens. Conceiving of the world as a single, great, indivisible soul, as in Plato's *Timaeus*, he asserts that the current of the cosmic spirit, the *spiritus mundi* constantly flowing in and around the sublunary world, must have a constant and powerful influence. The cosmic spirit is not compounded of the four earthly elements (earth, water, air, and fire), as is the human soul, but is rather constituted of a fifth element, the pure and incorruptible heavenly ether.

Nonetheless, there is an essential congruity between the human soul and the cosmic spirit, which is the basis of all sympathetic magic. The art of magic is to discover which substances and actions will promote and channel the influence of the cosmic spirit. The adept can accomplish this by consuming or otherwise associating himself with substances that have a powerful concentration of pure cosmic spirit, such as gold, wine, very white sugar, and the aroma of cinnamon or roses. To invoke the influence of a certain planet he may use the substances associated with it, or he may construct talismans. For the specifics, the adept is directed to the divine inspiration of the hermetica, which are filled with catalogues detailing which planets are governed by which stones, plants, animals, colors, parts of the body, and so forth.

Yet Ficino's preferred method of calling down the cosmic spirit seems to be the use of music. He makes a direct connection between the way that sympathetic plants, minerals, and other substances exert their influence and the principle of musical magic:

> [Just as] from a certain combination of herbs and vapors, made by Medical and Astronomical art, results a certain form, like a kind of harmony endowed with gifts of the stars; thus, from tones chosen by the rule of the stars, and then combined in accordance with the stars' mutual correspondences, a sort of common form can be made, and in this certain celestial virtue will arise. It is indeed very difficult to judge what kind of tones will best fit what kind

of stars, and what combinations of tones agree best with what stars and their aspects.

Difficult, but, you may be sure, not too difficult for Ficino: "Partly by our own diligence, and partly by divine destiny," he reports, "we have been able to accomplish this."

Saturn and Mars, malignant influences, do not have music but rather voices, the former "slow, deep, hoarse, and complaining," the latter "rapid, piercing, harsh, and threatening." The moon, being a neutral influence, has a neutral voice. The beneficent planets are described in this way: "Jupiter: music which is grave, earnest, sweet, and joyful with stability. Venus: music which is voluptuous with wantonness and softness. Apollo [the sun]: music which is venerable, simple, and earnest, united with grace and smoothness. Mercury: music which is somewhat less serious [than the Apollonian] because of its gaiety, yet vigorous and various."

Such characterizations of the music of the spheres have remained stock astrological lore right up to our own era; the modern concert-hall favorite *The Planets*, by Gustav Holst, is almost a literal rendering of Ficino's sketches of the planetary music. There can be little doubt that Ficino's own compositions must have been astrological music addressed to the sun, the planet of Apollo, the god of music. That planet is the one governed directly by musical influences, and in Ficino's scheme man himself is viewed as being primarily solar. His hymns to the sun were attempts to imitate the Pythagoreans, and he played them upon what he called an Orphic lyre.

Ficino's astrological music was based solidly in the Boethian notion of the essential congruity between the music of the spheres, *musica mundana*, the music of the human organism, *musica humana*, and ordinary music making, *musica instrumentalis*. That congruity, an aspect of the sympathetic relationship that exists between the human soul and the cosmic spirit (and which is reflected in the perfect proportions of instrumental music) causes like bodies to act in concert—in just the way that an open string of a lute, if it is in harmony with another string that is plucked, will vibrate sympathetically. This affinity, Ficino tells us, also explains why the human organism responds so profoundly to music. When *musica instrumentalis* is agitated, the soul of the person who hears it becomes agitated; when a composition is tranquil, the soul is tranquillized. An exactly parallel relationship exists between man and the cosmos: when a person's soul is in tune with the heavens, it responds just as sensitively to the music of the spheres. Thus, for example, if Mars is the dominant planet, the "music" of the soul will be harsh and martial; if it is Venus, then the soul will be voluptuous and delight in love.

Man, therefore, occupies the mean position, intermediary between the sublunary world and the cosmic spheres, which were governed by pure number for Ficino just as absolutely as for Pythagoras. In order to account for the powerful sympathetic vibrations that resonate between the three realms of musical-numerical proportion, Ficino invented a lovely and uncommonly satisfying theory: he conceives of musical concerts as possessing a living spirit analogous to that of the human soul and the cosmic spirit. His description of song is a delightful conceit: it is "warm air, even breathing, and in a measure living, made up of articulated limbs, like an animal, not only bearing movement and emotion but even signification, like a mind, so that it can be said to be, as it were, a kind of aerial and rational animal." A certain resemblance may be traced between the Ficinian song-animal and the man-made gods of Hermes Trismegistus's *Asclepius*—marvelous, magical creations motivated by an infusion of divine afflatus.

None of Ficino's own compositions have survived, but they must have been based directly upon the Orphic tradition, or, more properly, Ficino's conception of the Orphic tradition. We have a good clue as to what that conception might have been from a passage by Pico della Mirandola, an associate of Ficino's at Cosimo's academy who would carry Ficino's musical studies even further, into the realm of cabalistic magic. Pico wrote, "In natural magic, nothing is more efficacious than the Hymns of Orpheus, if there be applied to them suitable music, and disposition of soul, and the other circumstances known to the wise." As we saw in an earlier chapter of this book, Orpheus was considered to be a historical figure just as certainly as was Pythagoras or Plato (or, for that matter, Hermes Trismegistus). Thus Ficino saw himself as a scholar seeking to recreate an ancient musical tradition—just as Vincenzo Galilei and the other members of Bardi's Camerata were seeking to revive the practices of Attic drama (and not entirely unlike the musicians in our own day who are performing early music on instruments from the composer's lifetime).

The great difference between the first operas and the Orphic hymns of Ficino and Pico is exactly the difference between the secular drama and the sacred songs of antiquity upon which they were modelled. Ficino's solar hymns were probably closer to religious rites than musical concerts. The intent, certainly, was spiritual enlightenment, to bring the human soul into harmony with the cosmos; sensual pleasure of the sort that would be provided by the opera in the next generation was far from the minds, or at least from the avowed purpose, of the performers. Ficino's Apollonian concerts might have sounded something like a heroic ode declaimed by a young boy in honor of Pietro de' Medici, described by Poliziano in a letter to Pico della Mirandola: "His voice was neither like someone reading nor like someone singing, but such that you heard both, yet neither separately; it was varied, however, as the words demanded, either even or modulated, now punctuated, now flowing, now exalted, now subdued, now relaxed, now tense, now slow, now hastening, always pure, always clear, always sweet."

Those who followed Ficino on his quasi-religious quest abandoned any claim to being scientists. The unity of scientific and spiritual knowledge—the idea that to investigate the workings of this world was to seek to understand the order of the whole cosmos, and thus to glorify its Maker—was unraveling. Another indication of that may be found in the writings of Zarlino and Vincenzo Galilei, who make only perfunctory references to the notion of cosmic harmony. They were crossing the threshold that leads to modern science. It is true, as we have seen, that they were fallible and indeed prone to wild lapses, but their basic attitude was to seek after truth and let the chips fall where they may. (As Vincenzo's son Galileo would discover, the chips sometimes fell disastrously.) When the new breed of scientifically motivated *musici* did make mention of the spirit, they were probably thinking of something different from what Ficino and the Hermetic humanists meant by it; to them the term meant nothing more than the soul as defined by conventional Christian theology, or, even more prosaic, the medical sense of the word, as defined by Aristotle.

Although the Hermetic tradition had always operated in secrecy, it was nonetheless tolerated and even patronized by the church in the free and easy days of the Medici and Borgia popes. In Florence, Ficino himself was elected canon of the duomo, and when the famous engraved pavement of the duomo of Siena was laid down in the 1480s, one of the places of honor in this synopsis of human history was given to Hermes Trismegistus. In 1492, when the throne of St. Peter was vacated by the death of Innocent VIII, a rich Spanish cardinal named Rodrigo Borgia was chosen. He was crowned Alexander VI—a pagan name for a pagan pope.

In the first year of his pontificate he issued a full absolution of Pico della Mirandola, who had been under continual attack for heresy. It is easy for us now to see why Pico had trouble with the ecclesiastical authorities. While his oration entitled *On the Dignity of Man* is one of the earliest promulgations of the concept of free will, and a fundamental document of Renaissance humanism, other writings by Pico, notably the *Conclusiones magicae*, are unapologetic tributes to the power of magic. Pico's stated purpose is the familiar one of reconciling Christianity with pagan philosophy, but his sympathies at times lean decidedly toward the pagan end of the spectrum. Reading Pico, one comes away with the impression that Jesus was little more than a very successful and well-connected wizard. One of Pico's "magical conclusions" makes this astonishing heterodox assertion: "Nothing we know makes us more certain about the divinity of Christ than magic and Cabala."

Astonishing, but what is even more astonishing is that the Holy See should come to the rescue of the author. Yet Pope Alexander, the father of Cesare and Lucrezia Borgia, was far too fond of temporal power to pass up a possible means of gaining it simply because of a theological snag. In addition to his patronage of Pico the Pope also commissioned Pinturicchio to cover the walls of several rooms at the Vatican, now known as the Appartamento Borgia, with paintings in the "Egyptian" style, which glorify the power of Hermes and Hermetic magic. In one remarkable frieze, in the Room of the Saints, the bull symbolizing the Borgia family is identified with the sacred Apis bull of Osiris, the sun god, which is depicted worshipping the cross. It is hardly surprising that a pope who would openly identify himself with a pagan god would be sympathetic to magicians such as Marsilio Ficino and Pico della Mirandola, and even less surprising, perhaps, that the pendulum should swing back, that the ecclesiastical establishment ultimately rejected such blatant paganism and returned to a more orthodox theology.

The pontificate of Alexander VI marked the high-water mark of official acceptance of Hermetism. In the first years of the next century, Martin Luther began the process of reform, and by the end of the century, the Reformation and especially the Counter-Reformation were exerting a profound chilling influence on the intellectual climate throughout Europe, particularly in the occult arts. The year 1600 brought the first opera performance, and it was also the year that Giordano Bruno was burned at the stake. It was long believed that he was executed for his heretical advocacy of the Copernican system; but Frances Yates argues persuasively that while Bruno was immensely important to the history of cosmology, being the first to articulate the concept of the infinite universe, he burned not because he advocated heliocentricity, exactly, but rather because he actually practiced sun *worshipping*, among other forms of pagan mystery religion.

If Ficino and Pico describe a rising arc of Hermetic heterodoxy, then Bruno's coordinates fall off the edge of the chart. Rather than seeing in the ancient writings of Hermes Trismegistus a mystical prophecy of the coming of Jesus, Bruno believed that they were the pure expression of the true religion not only of this world but of an infinitude of worlds. In his lectures at Oxford he embraced such thoroughly un-Christian notions as metempsychosis, and came within a hair of rejecting Christianity outright.

It was the Copernican system of heliocentricity that got both Galileo and Bruno into trouble with the ecclesiastical authorities, but for entirely different reasons. Galileo realized that the earth moved around the sun because he was a good astronomer; Bruno preached heliocentricity because he venerated the sun and the notion of an eternally cycling earth suited his metaphysics. It will be recalled that heliocentricity had already made its debut in Western thought two thousand years before Copernicus, in the philosophy of Pythagoras himself. Of course for Pythagoras the concept was at once good science and good theology, for the two were the same thing.

The most radical expression of heliocentrism and the pagan revival in the Renaissance is embodied by the philosopher and rabble-rouser Tommaso Campanella. A renegade Dominican friar with a powerful, charismatic personality, Campanella envisioned a sun-worshipping Utopia, with himself as the high priest, and devoted his life to trying to bring it about. In 1599, he incited a revolt in Calabria against the Spanish monarchy, which then ruled southern Italy. The revolution, which was better equipped with ideals than practical notions about implementation, was easily quelled, and Campanella was imprisoned in Naples. He was tortured there and avoided being burned at the stake only because he cleverly feigned madness. During his twenty-seven years of confinement, Campanella wrote several books explaining his Utopian vision.

In his treatise *La città del sole*, Campanella describes the solar city, a schematic representation of the Pythagorean cosmos. In the center is a vast, perfectly round domed temple to the sun, hung with seven lamps representing the planets, which illuminate a map of the cosmos. The city is divided into seven concentric rings, also symbolic of the heavenly spheres, whose walls were covered with magical hieroglyphs and designs. The city is ruled by an all-powerful high priest, assisted by three subalterns: Power, who has charge of the military; Wisdom, the chief scientist; and Love, who uses astrology to determine who should procreate with whom, and when, in order to produce the happiest and luckiest race of people. There is much of the *Republic* and Thomas More's *Utopia* in the *Città del sole*, and Frances Yates also discerns the influence of the *Corpus Hermeticum* (part of the *Asclepius* is devoted to a description of an ideal city called Adocentyn, which is similar in many ways to the City of the Sun).

Strangely, the one field in which Campanella rejected orthodox Pythagoreanism was music: "In vain do Plato and Pythagoras make up a Music of the World out of our music; indeed they are talking nonsense. . . . If there is a harmony in the heavens and in the angels, it is of different order and has consonances other than the fifth, fourth, and octave." Campanella's understanding of Pythagorean music theory was obviously faulty; the music of the spheres was not based on "our music" but rather the other way round. Nevertheless, he accepts the idea that a celestial music does exist, but we are simply incapable of hearing it. Some day, he says, there will be new instruments that will allow us to hear the cosmic music, just as the telescope enabled man to see heavenly bodies that were invisible before.

After the execution of Giordano Bruno and the imprisonment of Tommaso Campanella, Hermetism slid ever deeper into official disfavor. The final blow came in 1614, when the Protestant humanist Isaac Casaubon proved that the *Corpus Hermeticum* was nowhere near as old as had been supposed, but was in fact a post-Christian fraud. Casaubon carried out a devastating textual analysis, showing just where the author or authors of the *Corpus* lifted which passages from the *Timaeus*, Genesis, the Gospel of John, and the writings of early church fathers such as Justin Martyr, Gregory of Nazianzus, and others. Most damning of all, Casaubon pointed out anachronistic references in the *Corpus* to Phidias and the Pythian games, who lived and which were played more than a thousand years after the time Hermes Trismegistus was supposed to have been writing—that, it would seem, stretched even the greatest power of prophecy beyond the acceptable limit. Reading the Hermetic texts today, it is difficult to understand how minds as acute as Ficino and Pico della Mirandola could have been taken in; the affair serves as an excellent parable of the power of suggestion.

As persuasive as Casaubon's debunking of the *Corpus Hermeticum* was, Hermes Trismegistus continued to be championed by zealous Hermetical scholars until the end of the seventeenth century. Foremost among them was Robert Fludd, an English doctor and occult philosopher who was one of the last examples of what came to be known as the "Renaissance man," although, born ten years after the death of Michelangelo, in 1574, he was himself something of an anachronism, who attempted in his works to put together again the Humpty Dumpty fragments of science, philosophy, and religion, which by then had been irrevocably sundered.

Fludd took as his subject everything. In a series of vast treatises that were among the most beautifully illustrated books of the seventeenth century, he rationalized (if the word can be used to describe so mystical a thinker) every aspect of the universe, visible and invisible, in a system that was essentially an expansion of the concepts of *musica mundana* and *musica humana* (although he never uses those terms except in a recognizably Boethian manner). Liberally drawing upon the authority of Hermes Trismegistus, Fludd asserts an underlying harmony and congruity between the universe, which he calls the macrocosm, and man, the microcosm. One of his principal metaphors for expressing this conceptual arrangement is *musica instrumentalis*. In an illustration in the first volume of *The History of the Macrocosm*, Fludd summarizes his cosmology in a figure he calls the Divine Monochord. A Pythagorean monochord comprising two octaves is divided into all the basic harmonic intervals, each of which describes an element of the universe. The scheme begins with low G, which is the earth, ascending to middle C, at which point God makes his appearance, and thence upward to high G, which is the most exalted division of the empyrean. The double octave of the whole chord represents the harmony of the universe, *musica mundana*.

Few of the individual elements of Fludd's system were original; all he was doing, essentially, was to reimagine in graphic terms the Cabalistic-Hermetic vision of Pico della Mirandola and overlaying it with the traditional notion of the Great Chain of Being. Fludd's unique contribution is the way he wedges all of human history into the cosmic scheme. In a treatise called *The Ape of Nature* (*De naturae simia*), which he characterizes as a technical history of the microcosm, he endeavors to show how all human arts and sciences are derived from number, just as the celestial and musical consonances, with which they have a mystical congruency, reflect the harmony between the human and celestial realms. Thus military science, cartography, mechanical engineering—as well as such "inner-

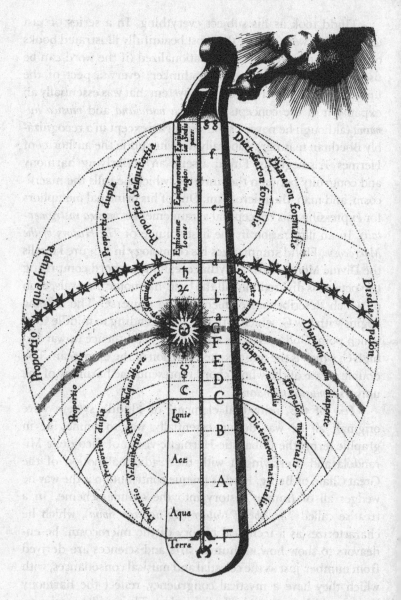

The Divine Monochord, by Robert Fludd

directed" human arts as prophecy, palm reading, geomancy, and the art of memory—have their cosmic counterparts. Fludd also draws the terrestrial elements and the planetary spheres into his great web of numerical-musical correspondences—a hugely ambitious undertaking that was to lead him into a bitter dispute with Johannes Kepler, his contemporary, who attempted to do the same thing with magic-tinged mathematics, as opposed to Fludd's vaguely mathematical species of magic.

Fludd was especially devoted to the art of memory, an important technique for the secret societies such as the Fraternity of the Rosy Cross, the Rosicrucians, of which he was a member. Some of the most beautiful illustrations in his books are memory palaces, mnemonic aids that enabled adepts to memorize whole volumes of philosophy and learning, an accomplishment that would obviously be of great use to a man of Fludd's encyclopaedic interests. One such illustration, "The Temple of Music," is a memory palace devoted to *musica instrumentalis*. This engraving, based upon the stage architecture of Baroque theatres, summarizes the basics of music theory as they would have been known to a rather old-fashioned pedant of the early seventeenth century. On the left side is the Divine Monochord, and on the porch next to it the scene of Pythagoras discovering the harmonic proportions at the blacksmith's shop. Above this vignette are Crantor's cosmogonic lambda in explication of the *Timaeus* (although the engraver has got the numbers wrong) and a checkerboard scheme that functions as an aid to musical composition, showing which notes harmonize with each other. A clock illustrates the relative time values of the notes, and the columns and rows of bricks teach the scales. Finally, a Muse stands in an alcove, pointing to a model composition, rather disappointing from a musical point of view, based upon the principles inculcated by the temple. It is an admirable accomplishment on its own terms, but to Fludd's contemporaries, who were listening by now to the technically sophisticated operas of Monteverdi, the erotic lute songs of John Dowland,

"The Temple of Music," by Robert Fludd

and the keyboard fantasies of Orlando Gibbons and Girolamo Frescobaldi, the polyphonic orthodoxy of "The Temple of Music" must have seemed as quaint as Grandma's bloomers.

Most of the graphic designs in Fludd's books are based, in some cases covertly and in other cases explicitly, upon magic wheels. In these mandala-like charts, the congruities between celestial and earthly harmonies are explicated as plainly as the engraver's art permitted. For Fludd, everything that existed, or that could be thought of, had its place not only in this world but also a parallel, by virtue of mystical kinship, in every other realm. A considerable amount of shoehorning and fudging was required to fit everything into place, but that, it would seem, was just what magic was for. For sheer grandiosity of vision and its spectacular intellectual nerve, Fludd's system towers alone, albeit in a slightly mad way, like Aristotle in Wonderland.

The last practitioner of speculative music, in the classic, Pythagorean sense, was a Geneva-born Jesuit antiquarian named Athanasius Kircher. Whereas Robert Fludd was a dyed-in-the-wool devotee of Hermetism, Kircher, a churchman who carried out most of his work in Rome in the era immediately following the execution of Giordano Bruno and actually during the trial and condemnation of Galileo, had to be very careful about not straying into theological error. The Society of Jesus, in its way almost as mysterious and aloof an organization as the Rosicrucians, gave Kircher a measure of independence and therefore intellectual freedom, but his greatest protection was afforded him by the extreme erudition of his works.

Kircher might be described as a Baroque proto-anthropologist, a collector and systematizer of every conceivable type of cultural expression from around the world. It was his passionate conviction that all the religions of the earth, ancient and modern (with the notable exceptions of Islam and the Gnostic heresies), were aspects of the one true religion, the familiar syncretism of cabalistic-Hermetic mysticism and Christianity, with the former as usual miraculously presaging the latter. Jesuit missionaries returning to Rome from posts throughout the world brought him every kind of literary and artistic curiosity, all of which he proposed to comprehend in his great cultural synthesis.

Many of Kircher's writings now seem fanciful to the point of being bizarre: a natural history of dragons and basilisks; a comparative study of giants; a map of Atlantis; and a number of Rube Goldberg-like inventions, including a talking statue and a botanical clock (consisting of a sunflower on a piece of cork, floating in a dish of water, which rotates as it follows the sun). Yet other inventions of his proved to be useful, such as the magic lantern and the megaphone. Kircher was also a pioneer in the development of techniques of artistic perspective; Nicolas Poussin was among his pupils in that field. His collection of artifacts, the Museo Kircheriano in Rome, was, with the alchemist Elias Ashmole's antiquary treasure house at Oxford, one of the first public museums in Europe.

Like Fludd, Kircher was an aficionado of all things Egyptian and a firm believer in the antiquity of Hermes Trismegistus, whom he placed as a contemporary of Abraham (a rather strange belief, it might seem, nearly half a century after Casaubon's debunking of the *Corpus Hermeticum*). He wrote a number of fat books purporting to decipher the hieroglyphs of ancient Egypt. Kircher's "translations" were pure invention; he discovered sublime Neoplatonic hymns in the texts of obelisks that, 170 years later, when Champollion translated the Rosetta Stone, were revealed to be nothing more than a boring list of rulers and their conquests. In one treatise on the subject, the *Oedipus Aegyptiacus* (the strange title presumably refers to the riddling nature of the hieroglyphs), Kircher indulges in a final burst of enthusiastic Hermetism, enumerating for the last time in a work of any intellectual respectability the Ficinian catalogue of the great teachers of divine wisdom, commencing with Hermes, followed by Orpheus, Pythagoras, Plato, and the others.

As in any Neoplatonic system, music played an essential

part in Kircher's philosophy. In a curious little book called *The Ecstatic Journey* (*Iter exstaticum*), which is clearly modelled on Plato's Myth of Er and *Scipio's Dream*, he describes an excursion through the celestial spheres. After listening to a concert by three lutenists, the protagonist, Theodidactus (God-taught), lapses into an ecstatic trance and is led by a spirit named Cosmiel on a journey through the crystal spheres of heaven. Kircher's cosmology is based upon that of Tycho Brahe, the semi-Copernican view that was the "politically correct" system in Rome after the condemnation of Galileo. In the Tychonian universe, the earth was the center, as theology required, but the other planets and the stars revolved around the sun, as astronomy required; the solar system itself revolved around the earth. Kircher's star traveller visits each of the planets, which are uninhabited but governed by seraphic Intelligences, hears the music of the spheres, and, after he reaches the outermost sphere of the fixed stars, returns to earth.

Kircher's musical summa is an encyclopaedic volume called the *Musurgia universalis*, published in 1650. More than most of Kircher's writings, it is a rather up-to-date work, covering the new Baroque musical styles and the still-emerging opera, as well as the older polyphonic music. According to the modern musicologist Joscelyn Godwin, Kircher was the first to articulate "the Baroque 'doctrine of the affections,' according to which the purpose of music was to illustrate or imitate various emotional or affective states: a doctrine that was the very basis of early opera, as well as one of the unquestioned assumptions of later composers such as Bach and Handel."

Always searching for the first causes of things, in the *Musurgia* Kircher devotes a great deal of speculation to the subject of Greek music. He records what he claims is the transcription of an ancient setting of an ode by Pindar he found in a Sicilian monastery, although no one afterward was ever able to locate the manuscript. Kircher has been accused of forging the piece, but Godwin is probably right in asserting that he was only guilty

The Musarithmetic Ark, composing machine attributed to Athanasius Kircher

of "the more characteristic faults of misunderstanding and lack of criticism."

Another musical enigma involving Kircher is presented by a composing machine called the "musarithmetic ark," which is now in the collection of the Pepysian Library at Magdalene College, Cambridge. Based upon the principle of Fludd's composing chart in "The Temple of Music," and in its operation somewhat similar to Napier's Bones, the calculating apparatus invented by the Scottish mathematician John Napier, this composing machine was, according to Joscelyn Godwin, the invention of Athanasius Kircher. We know that Samuel Pepys owned the *Musurgia universalis*, for he wrote in his diary in 1667 that he bought a copy of it. However, the present librarian of the Pepysian Library, Richard Luckett, maintains that the musarithmetic ark was never in Samuel Pepys's possession, nor does it have any connection whatsoever with Athanasius Kircher; for him it is a mysterious piece of learned flotsam, washed up onto the shelves of the Pepysian Library at an unknown time, from an unknown source.

Kircher's passion for categorization reaches its peak in a chart called the "Enneachord of Nature," a system that encompasses every aspect of the phenomenal world in magical-musical terms. An enneachord is a nine-stringed instrument; Kircher describes ten of them, each attuned to a different class of objects or qualities. The principal enneachord is tuned according to the ancient Greek system of harmonic proportions, which are identified in the usual way with the planets and the fixed stars. Next to it, in the first position, is the enneachord devoted to the hierarchy of heaven, with God occupying the highest position. The other enneachords depict the harmonious (and disharmonious) relationships among mineral substances, stones, plants, trees, aquatic creatures, birds, quadrupeds, and colors.

If the string of Saturn is struck, then all of the saturnine things (lead, topaz, hellebore, cypress, etc.) will vibrate sympathetically. Moreover, since the harmony (or disharmony) of the musical intervals is well known, we may determine the sympathetic relationships among all of the other substances in the chart. The earth and Saturn form an octave on the enneachord of the planets, which means that earthly things and saturnine things are in sympathetic harmony. On the other hand, according to Kircher, since Jupiter and Saturn are out of tune with each other, "the things of Jupiter are quite out of tune with those of Saturn on account of their dissonant interval. . . ." In other words, one ought never to mix amethysts, lemons, or roses with an owl, for the result will be disharmonious. Yet it is permissible to combine fruits, eels, and lodestones with owls, for they are in consonance with the earth, which is in harmony with Saturn and saturnine things.

Such a synopsis will, of course, make the system look absurd; nor is it altogether lacking in absurdity. Yet it was not meant to be taken literally. Kircher himself says, "No one can explain this symphony in verbal concepts, none declare it with however felicitous or eloquent a pen, none can penetrate it with the profoundest scrutiny of mind." In a prayer to the Great Harmost (that is, God in his incarnation as the tuner of men's souls), Kircher petitions that he "tune the enneachord of my soul to thy divine will; play upon all the strings of my soul to the praise and glory of thy name, that I will love thee with Seraphic ardor, and seek thee constantly with a Cherubic mind."

In the post-Lutheran era this is rather suspect theology: if Kircher had been a rural parish priest in Central Europe during those witch-hunting days, rather than a member of the Society of Jesus, safely ensconced behind the walls of the Vatican, it is doubtful how long he would have been permitted to carry on such fanciful talk. Yet on the other hand, for a man thirty years younger than Johannes Kepler it is intellectually jejune. Mystical visions of the cosmos would continue to flourish with undiminished vigor down to the present day, of course, even as scientific investigation became ever more "pure" and

sophisticated. Yet Kircher was the last Western thinker who
believed that he could do both, who managed to keep himself
oblivious to the widening psychosis in Western thought. In
some ways his greatest accomplishment was simply his boundless self-confidence in an age in which doubt and scientific
uncertainty had taken hold. Johannes Kepler, although a better
mind and vastly more influential, was attempting to do funda-
mentally the same thing. The difference is that Kepler knew
how hopeless it was; even as he expended more and more
amazing brainpower in trying to resolve the conflict between
classical wisdom and scientific investigation, he veered toward
madness.

Kepler Pythagorizes

With Kepler, we are in the presence of someone who heard the music of the spheres not as a noble echo from antiquity but as a palpable symphony in his ears, as Pythagoras had done. He contemplated the heavens with the profound awe of a good Platonist, yet somehow also with the clear-eyed, unsentimental gaze of a modern scientist. In him a great intellect and a great soul struggled courageously to coexist. Arthur Koestler calls Kepler the watershed; to choose another aqueous metaphor, he might be likened to the dike separating the placid waters of the classical age and its renaissance from the first flow of the treacherous waters we may finally call the modern age. Strange that the rule of reason should bring chaos into the world, but that is the conclusion to which the history of science brings us time and again.

Kepler's masterpiece, the fifth book of *The Harmony of the Universe* (*De harmonice mundi*), is the summa of the great theme, the supreme treatise on the musical universe. It is a great work of modern science, yet it could only have been written by someone who, like every other intellectual of his age, sincerely believed in the efficacy of magic, the essential truth of alchemy, the authenticity of the *Corpus Hermeticum* and the Cabala, and the influence of the stars on human affairs. It would be easy to make too much of this. A modern enthusiast of the esoteric arts, hostile to the fact-bound literalness of orthodox science, might claim that this duality in Kepler's thought somehow diminishes the legitimacy of "real" science: "Here's one of your principal heroes," the hypothetical Hermetist might say, "and yet while he was making all of his greatest discoveries he was a firm believer in astrology, alchemy and other so-called superstitions." That would be wrong, but it would be equally wrong to assume that Kepler was somehow just lucky, that he stumbled upon a few laws of planetary motion while he was fooling around with his astrological charts. The usual way round this dilemma has been simply to disregard Kepler's esoteric leanings (something which can only be accomplished by not reading his books) and draping him in a rather ill-fitting modern scientist's white coat.

Kepler knew exactly what he was doing. In order to understand what he was about, we must try to imagine what life was like in an age when people still looked to the sky. Before there was smog and skyscrapers—that is, throughout all human history until about a hundred years ago—it was accepted as an indisputable truth that the heavens were of a transcendental importance that far outweighed the petty doings of this earth. It was assumed that what went on up there must have a connection with what happened down here; the question was, exactly what sort of connection was it? And people then being no different from what they are now, that raised the really interesting question of how we could get the cosmic drift and use it to influence human affairs. Kepler despised that petty, greedy sort of astrology. He thought that if we could only obtain a sufficient quantity of the right kind of astronomical observations, then we might be able to fashion a new astrology, one that was ethical and spiritual. In one of his many treatises about astrology, he wrote, "No man should hold it to be incredible that out of the astrologers' foolishness and blasphemies some useful and sacred knowledge may come, that out of the unclean slime may come a little snail or mussel or oyster or eel, all useful nourishments; that out of a heap of lowly worms may come a silk worm, and lastly that in the evil-smelling dung a busy hen may find a decent corn—nay, a pearl or golden corn—if she but scratches and scratches long enough."

Kepler explicitly acknowledged Pythagoras and Plato as his conceptual masters, and he adhered closely to their idealistic schema of a universe ruled by perfect, mathematical music. What distinguishes his vision of celestial harmony is that, for the first time, the music of the spheres is conceived as being polyphonic. For Plato and Cicero and the scholastics the music of the spheres was real enough (even if nobody one knew had ever heard it), yet it consisted only of scales. As we know, the classical notion of harmony was serial; the harmony is apprehended by the mind, as an abstract relationship between two tones rather than a single concord heard as a unity by the ear.

The discovery of polyphony was probably the single most important advance in theoretical music in the post-classical age, yet it was greeted by most of Kepler's contemporaries with a strange indifference. The antiquity-worshipping mania effectively precluded any chance that a modern innovation such as polyphony could ever be considered a great improvement over the music that preceded it, for that would imply that ancient music was inferior, an inconceivable notion. Yet Kepler, for all his veneration of Pythagoras, whom he called "the grandfather of all Copernicans," was very much a modernist and a believer in progress. He continually stresses that polyphony was unknown to the ancients (citing as his authority Galilei's *Dialogo della musica antica e della moderna*), and he asserts that his new astronomy will be as much of a great step forward in cosmology as polyphony was in music.

This notion of simultaneity was an essential component of the Keplerian universe. He was not especially interested in accounting for the motions of a few planets and stars, or a thousand of them; a good Hermetist at heart, Kepler wanted nothing less than to explain the very structure of the cosmos. His first book, published when he was a twenty-five-year-old professor of mathematics at the seminary at Graz, carried this overwhelming title: *An Introduction to Cosmographical Treatises, Containing the Cosmic Mystery of the Admirable Proportions between the Heavenly Orbits and the True and Proper Reasons for Their Numbers, Magnitudes, and Periodic Motions*—otherwise known as the *Mysterium cosmographicum*.

The book has a delightfully chatty tone, rather surprising in a work that sets out to solve the riddle of the cosmos. Kepler begins by stating that he accepts the Copernican model of the heliocentric universe. Then he tries to make sense of the planets, their distances and velocities; he wonders why there are just six of them, "instead of twenty or a hundred." Surely there was a reason for such a circumstance: Would Providence have disposed the elements of creation in a random or irrational way? Kepler's certainty that there was a rational and eternal plan in the cosmos drove everything in his thinking. To justify that conviction, he draws a characteristically grandiose analogy between the three realms of the universe and the three aspects of the Trinity: "I was made bold to attempt this," he writes in the *Mysterium cosmographicum*, "by the beautiful harmony that exists between the parts [of the cosmos] that are at rest, the sun, the fixed stars, and the intermediate space, and God the Father, the Son, and the Holy Ghost: a similarity I shall pursue further through cosmography. Since the parts that are at rest are disposed in this way, I did not doubt that the moving parts would also be harmonious."

Yet try as he might, Kepler could not find the underlying pattern in the planetary system. At first he tried inserting phantom planets between Mercury and Venus and then between Jupiter and Mars, in an effort to rationalize the planetary intervals, but to no avail. Then he had a glorious revelation: "I believe that Divine Providence arranged matters in such a way that what I could not obtain with all my efforts was given to me through chance; I believe all the more that this is so as I have always prayed to God that He should make my plan succeed, if what Copernicus had said was the truth." On July 9, 1595 (he exultingly provides us with the date of this momentous event), while he was drawing a figure on the blackboard for a plane geometry class, he had a sudden flash of insight. The figure was a triangle with a circle inscribed within it, which was itself inscribed within a circle, thus:

What struck him was that the ratio between the two circles was congruent with the ratio between the orbits of Saturn and Jupiter. Moreover, those two were the outermost spheres; he calls them the first planets, "and the triangle is the first figure in geometry. Immediately I tried to inscribe into the next interval between Jupiter and Mars a square, between Mars and the earth a pentagon, between the earth and Venus a hexagon," and so forth.

Divine inspiration had to lead him to take one more step, a leap into the third dimension. The ratios between the heavenly spheres cannot be described as plane figures: the planets' revolutions are accomplished in three dimensions—the heavenly spheres themselves are three-dimensional—so it follows inevitably that the ratios between them are described by solid figures. "Why look for two-dimensional forms to fit orbits in space? One has to look for three-dimensional forms—and behold, dear reader, now you have my discovery in your hands!"

When he began to try out his theory, he was struck by the

ultimate thunderbolt: the number of perfect solids is five, just the number of figures that would be necessary to describe the intervals between the planetary spheres. The perfect solids, also known, appropriately enough, as the Pythagorean solids and the Platonic solids, are so called because they are perfectly symmetrical; their faces are all regular polygons of the same shape and size. It is a fact of geometry, proved by Euclid, that only five solids fit this description: the tetrahedron (pyramid), the cube, the octahedron (faced with eight equilateral triangles), the dodecahedron (twelve pentagons), and the icosahedron (twenty equilateral triangles):

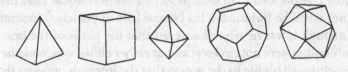

Like the figure he had drawn on the blackboard for the seminarians in Graz, the perfect solids may have a sphere circumscribed around them, touching at every vertex, and a sphere may likewise be inscribed within them. The circumsphere and the insphere, as they are called, are also symmetrical and concentric. These polyhedrons were, for Kepler, the most beautiful and perfect because they most nearly imitated the sphere, which in the *Timaeus* was identified as an image of God, a concept Kepler accepted as an article of faith. When Kepler compared the insphere-to-circumsphere ratios of the five Pythagorean solids, which in his emerging scheme would locate the heavenly spheres in space around the sun, they seemed to match with the ratios of the planetary motions. The young mathematician leapt joyously to the conclusion that he had found the key to the universe:

> It is amazing! Although I had as yet no clear idea of the order in which the perfect solids had to be arranged, I nevertheless succeeded . . . in arranging them so happily,

that later on, when I checked the matter over, I had nothing to alter. . . . Within a few days everything fell into its
place. I saw one symmetrical solid after the other fit in so
precisely between the appropriate orbits, that if a peasant
were to ask you on what kind of a hook the heavens are
fastened so they do not fall down, it will be easy for you to
answer him.

The resulting system, so startling in its perfection (at least
to its author), summarized the whole cosmos in a single complex
figure, which Kepler had engraved and folded, into the *Mysterium cosmographicum*. Of course, the motions of the planets are
more complicated than that, as Kepler himself would later discover when he formulated his famous laws of planetary motion.
Although it was he who discovered that the paths of the heavenly bodies were not circles at all but rather ellipses, he nonetheless clung all his life to the notion that the intervals between the
celestial spheres were described by the five perfect solids.
Twenty-five years later, when he published a second edition of
the *Mysterium cosmographicum*, he said of his youthful work that
"never before has anybody published a more significant, hap-
pier, and in view of its subject, worthier first book." As Arthur
Koestler suggests, this is not mere braggadocio but rather the
certitude of the mad, for Kepler, like many another person of
unquestioned genius, was a little demented.

This book is not the place to summarize Kepler's great and
original accomplishments in astronomy, which really begins in
its modern form in his work. Yet his first book, the *Mysterium
cosmographicum*, and his greatest book, *The Harmony of the Universe*, both deal with the structure of the cosmos in a way that is
classically Pythagorean. While the *Mysterium cosmographicum*
does not explicitly deal with the musical aspect of the heavenly
spheres, it nonetheless sets forth the bare outlines that, in *The
Harmony of the Universe*, will resound with music. By the time he
came to compose his big book, Kepler had become the most
famous astronomer in Europe and Mathematicus to the insane

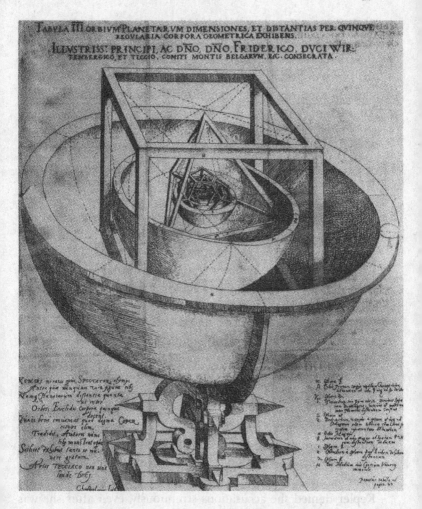

Frontispiece to Johannes Kepler's *Mysterium cosmographicum*

Holy Roman Emperor Rudolf II, a post that seemed to consist almost entirely of casting horoscopes. John Donne sarcastically described Kepler during this period of his life as a man "who (as himselfe testifies of himselfe) . . . hath received it into his care, that no new thing should be done in heaven without his knowledge."

Then disasters began to pile upon him. In 1612 Rudolf died, leaving Kepler without court protection. As he was readying *The Harmony of the Universe* for the press, his beloved daughter died. He was excommunicated. War was imminent. But most traumatic of all, his mother, Katherine Kepler, was charged with witchcraft and narrowly avoided being burned at the stake. The first years of the seventeenth century were the height of the witch-hunting mania in Germany. The little town of Leonberg, where she lived, had no more than two hundred families, yet in the winter of 1615 alone six witches were burned. Seventy years old, described by her son in the horoscope he cast for her as "small, thin, swarthy, gossiping and quarrelsome, of a bad disposition," and probably a trifle addled as well, Frau Kepler was an ideal target for the witch hunters. The aunt who had raised her had been found guilty of intercourse with Satan and had burned at the stake. One of Katherine Kepler's neighbors accused "Little Kate" of sickening her with a magic potion. The sexton of a nearby town testified that Frau Kepler had asked him to give her her father's skull, so that she could have it mounted in silver and made into a drinking cup for her son Johannes, the astronomer—grisly, but it was a grisly age. Frau Kepler denied the accusations strenuously, even after she was subjected to the usual viewing of the instruments of torture, but the more she swore her innocence, the more surely she seemed to be doomed. Kepler wrote hundreds of passionate pages in her defense. In the end, she was released after fourteen months of imprisonment, uncondemned yet unexonerated. She could not return to her home in Leonberg lest she be lynched; six months later she was dead.

It was under such conditions that Kepler wrote *The Harmony of the Universe*, and yet not since antiquity had there been a book of science that expressed itself with more sweetness, purity, and symmetry. Arthur Koestler calls it, neatly, "a mathematician's Song of Songs," and he scarcely overstates the case when he writes: "What Kepler attempted here is, simply, to bare the ultimate secret of the universe in an all-embracing synthesis of geometry, music, astrology, astronomy, and epistemology. It was the first attempt of this kind since Plato, and it is the last to our day. After Kepler, fragmentation of experience sets in again, science is divorced from religion, religion from art, substance from form, matter from mind." That process of fragmentation had in fact already begun; Kepler was trying to hold together what had already flown apart. Standing astride that line where the last pale gleams of antiquity joined with the first stirrings of modern rational thought, he was able to look in both directions. But after that moment passed, it would never again be possible.

By the time of *The Harmony of the Universe*, Neoplatonic notions of the crystal spheres had become encrusted with a poetic patina, yet for Kepler the spheres were absolutely real. As far as he was concerned, there was no doubt that they made music, but it was literally worlds apart from Marsilio Ficino's lute strumming and solar warbling. "The movements of the heavens," Kepler wrote, "are nothing except a certain everlasting polyphony, perceived by the intellect, not by the ear."

Despite his devotion to the modern notion of polyphony, Kepler keeps his conception of harmony purely mathematical. In his essay on Kepler's celestial music D. P. Walker gives a concise definition of the classical notion of harmony: "Harmony, musical or of any other kind, consists in the mind's recognizing and classing certain proportions between two or more continuous quantities by means of comparing them with archetypal geometric figures." For Kepler just as for Pythagoras, the major third, which is defined by the ratio 4:5, is not a pair of notes to be twanged on a lyre or plunked on the keyboard of a virginals, although those are valid expressions of it; rather it is a mathematical ideal, of a divine substance, that need not even be expressed in order to exist, for it is eternal. In one of his defenses of astrology, the *Tertius interveniens*, Kepler writes that God "has represented Himself in the world," and the way he has done so is through mathematics: "I sometimes wonder whether the whole of nature and all the beauty of the heavens is not symbolized in geometry."

At a critical juncture, however, Kepler parts company with the Pythagoreans. After they discovered by ear the perfect consonances of the octave, the fifth, and the fourth (respectively 2:1, 3:2, and 4:3), Kepler says that they turned away from the evidence of their ears and excluded other perfect geometric archetypes. "The Pythagoreans were so addicted to this kind of philosophizing in numbers," he wrote in the third book of *The Harmony of the Universe*, "that they failed to keep the judgment of their ears.... They defined solely by their numbers what is a harmonic interval and what is not, what is consonant and what is dissonant, thus doing violence to the natural instinctive judgment of the ear." He was thinking in particular of thirds and sixths (5:4 and 6:5) and the minor tone (10:9), which were excluded by the Pythagoreans but which are harmonious to "all well-eared musicians of today," as Kepler puts it—and, more to the point, they are essential to his theory. Thus while the harmony of the cosmos is perceived by the intellect, the ears could nonetheless be used to provide some clues as to essential parts of its mathematical superstructure.

Kepler's harmonic scheme is based firmly on standard Ptolemaic geometry. Using the ruler and compass, the only tools permitted in classical mathematics, a geometer may construct all the perfect polygons (triangle, square, pentagon, etc.); once he has done so, he will have a complete set of illustrations of the harmonic consonances. For example, if you inscribe a pentagon in a circle (which is how one is constructed in classical geometry), the ratio between the arc created by one side of the pentagon and the rest of the circle will be precisely the ratio of the major fifth, 4:5, thus:

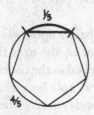

Conversely, all polygons that cannot be construed with compass and ruler are abominations: *inscibilia inefabilia nonentia*, unknowable, unmentionable non-beings. The example Kepler cites is the heptagon, which cannot be constructed with a ruler and compass. The ratios produced by a heptagon are 1:7 and 6:7, which are odious, non-musical intervals.

Much of *The Harmony of the Universe* is devoted to proving how the harmonious consonances may be derived from the perfect polygons, which rank with the perfect Pythagorean solids as the purest expressions of the geometer God. In the fifth book, Kepler undertakes to prove the relationship between the musical ratios and the motions of the planets—the music of the spheres. After his own discovery of the elliptical paths of the planets, the traditional circular orbits were no longer valid. He attempted any number of schemes to reconcile the musical ratios and the measurements of the planetary revolutions that he had at his disposal: he tried constructing a series based on the planets' periods of revolution, on their relative volumes, on their perihelia and aphelia (nearest and farthest distances from the sun), on their extreme velocities. He tried comparing the length of time a planet needed to traverse an arc of its orbit at aphelion with the time required to cover the same distance at perihelion, but that did not work either. Then he had another brainstorm: he compared the distances covered at perihelion and aphelion *as if they were being observed from the sun*—which is certainly the vantage point that would be taken by a Master Geometer in the Copernican cosmos—and the ratios thus derived fitted with thrilling precision the musical ratios of the perfect polygons.

Kepler's ratios are based upon what he calls the apparent diurnal movements, that is, the arcs that to a solar observer would appear to be travelled in the course of a twenty-four-hour period at perihelion and aphelion. For example, Saturn traverses an arc of 135 seconds per day when it is nearest the sun, if viewed from the sun. On the day it is farthest from the sun, the planet would appear to a solar observer to travel an arc of 106 seconds. The ratio of 135 to 106 is just a tiny fraction off 5:4, a major third. In fact it is a ratio of 5:3.9259, which is close enough to satisfy most ears (and, presumably, most intellects). Using this method, he found to his delight that all six planets produced ratios almost precisely equivalent to harmonies that were expressible by perfect polygons. Jupiter's perihelial/aphelial ratio was, more or less, 6:5, a minor third; that of Mars 3:2, a fifth; that of the earth 16:15, a semitone; that of Venus 24:25, barely equivalent to the Pythagorean comma; and that of Mercury 12:5, an octave and a minor third.

The little discrepancies did not perturb Kepler in the slightest. It would not be reasonable to expect the music of the spheres to correspond exactly with earthly instrumental music, for the one was in no way patterned after the other. Rather, they were both reflections of the same divine and eternal archetypes. A modern scientist would be quick to fall back on the position that such discrepancies were the result of inaccuracies in the measurements arising from the inherent limitations of the instruments used for making celestial observations. But that would never have occurred to Kepler, whose mind saw in every difficulty a challenge to be overcome. Nowhere in his writings does he cut corners. Rather, he devotes pages and pages of ingenious explanation in *The Harmony of the Universe* to disposing of all those troublesome little fractions.

The most marvelous revelation came when he began to make ratios by pairing off the planets: "But in the extreme movements of two planets compared with one another, the radiant sun of celestial harmony immediately breaks in all its clarity through the clouds." Using these extreme values, Kepler was able to construe the entire musical scale. Furthermore, he discovered that each of the planets has its own scale, which is also determined by its speed at perihelion and aphelion. For example, Saturn, the deepest, and Mercury, the highest pitched, would be notated as follows:

Kepler points out that the changes in tone would not be divided into steps, as required by staff notation, but would rather be an eternal note that rises and falls continuously, like a trombone player forever moving his valve back and forth, or a violinist pushing his finger up and down a string. In the case of the earth it is more like a vibrato, because the entire range of the "scale" amounts to scarcely more than a semitone; that of Venus is even smaller, spanning an interval of 25:24. Finally, Kepler sets out to construct a celestial motet, a chord to unite the songs of all the planets. Because of the tiny span of the scales of the earth and Venus, the possibilities are severely restricted. Kepler doubts whether such a heavenly symphony could have happened twice; more likely it occurred just once, at the moment of creation, and will only do so again on the Day of Judgment.

Thus it is no idle figure of speech when he states that "the movements of the heavens are nothing except a certain everlasting polyphony." One chapter of *The Harmony of the Universe* is devoted to the question posed by its title: "In the Celestial

Harmonies, which Planet Sings Soprano, which Alto, which Tenor, and which Bass?" The answer is, Mercury is the soprano, the earth and Venus share the alto part, Mars is the tenor, and bass is shared by Saturn and Jupiter. He justifies this fancy not only by the obvious method of their relative pitch but also for solid musical reasons. Mercury of all the planets is the freest and swiftest, making it most like a soprano. The narrow range of the earth and Venus, which are very near to each other in pitch, makes them suitable altos. "As the tenor is free, but nonetheless progresses with moderation, so Mars . . . can make the greatest interval, namely a perfect fifth." And since the bass, like the alto, had to be doubled, as it was in the musical performances of Kepler's day, the inaudibly deep scales of Saturn and Jupiter, just an octave apart, make them ideal for the part.

Kepler's great synthesis, certainly the most ambitious and comprehensive description of the music of the spheres ever attempted, was completed on May 27, 1618, just three days after the Defenestration of Prague sparked the Thirty Years War. Two years later, in a different context, Kepler wrote, "In vain does the God of War growl, snarl, roar, and try to interrupt with bombards, trumpets, and his whole tarantantaran. . . . Let us despise the barbaric neighings which echo through these noble lands, and awaken our understanding and longing for the harmonies."

The rational scheme of Kepler's universe provoked enormous controversy at the time of its publication. Chief among his adversaries was none other than Robert Fludd, the Rosicrucian doctor of Oxford. It was actually Kepler who started the feud, with a personal attack on Fludd in an appendix to the fifth book of *The Harmony of the Universe*. Taking a defensive posture, as if anticipating the criticism that might be levelled against his work, Kepler speaks to the Rosicrucian directly: "When I pronounce your enigmas (that is, your harmonies) obscure, I speak according to my judgment and understanding, and I have as an aid in this matter you yourself, since you deny that your purpose is subject to mathematical demonstration, without which I am like a blind man." Kepler, for whom Copernican heliocentricity was tantamount to a religious belief, was contemptuous of Fludd's geocentricity, which after all was based not on observation but on the humanistic tenet of the primacy of man's place in the universe.

Yet Robert Fludd was not one to shrink from a challenge. He responded in the second volume of *The History of the Macrocosm* with a vigorous denunciation of the vulgarity of Kepler's mathematics. Fludd considered himself to be among the select group of *sapientes*, those wise few who have been initiated into the eternal mathematics that govern creation. Kepler, unacquainted with the mysteries of alchemy and the Rosicrucian brotherhood, was thus hopelessly sidetracked with his endless carping about measurements. For Fludd, the very notion of measuring the universe was erroneous.

In one of his broadside attacks against Kepler, the *Demonstratio analytica*, Fludd explains the essence of his criticism: "For it is for the vulgar mathematicians to concern themselves with quantitative shadows; the alchemists and Hermetic philosophers, however, comprehend the true core of the natural bodies." Taking on Kepler directly, he says scornfully, "He puzzles out the exterior movements of the created thing, whereas I behold the internal and essential impulses that issue from nature herself; he has hold of the tail, whereas I grasp the head; I perceive the first cause, he the effects." Furthermore, even if Kepler is correct by the vulgar and even delusionary terms of his mathematics, "nevertheless he is stuck too fast in the filth and clay of the impossibility of his doctrine and, perplexed, is too firmly bound by concealed fetters to be able to free himself without damage to his honor. . . ."

One of the superficial points that the two men might have seemed to share in common was their use of graphic representations to propound their theories. Yet they both go to great pains to ensure that no one think for a moment that they are after the same thing. Kepler describes Fludd's system of interpenetrating pyramids as hieroglyphs and pictures, while his own diagrams, he says, are not there as ornaments to divert the reader but because he is doing mathematics. He points out that the ratios of the planetary spheres that Fludd uses in his Divine Monochord do not correctly reflect the empirical data. Fludd responds that *sapientes* are not in full agreement as to the correct ratios of the spheres, and that such minor quibbles do not matter in any case. Yet Kepler points out, quite rightly, that they are of essential importance if you want to make music. He himself has gone to such extraordinary lengths to reconcile the observed planetary motions with the Pythagorean ratios that he is understandably annoyed with a self-proclaimed Pythagorean who cannot even get the musical ratios right.

In a study of the Kepler-Fludd controversy, the great Viennese physicist Wolfgang Pauli advances the familiar image of the former as a great modern scientist trapped in a backward age of superstition. In the first half of the seventeenth century, he says, "the then new, quantitative, scientifically mathematical way of thinking collided with the alchemical tradition expressed in qualitative, symbolical pictures: the former represented by the productive, creative Kepler, always struggling for new modes of expression, the latter by the epigone Fludd who could not help but feel clearly the threat to his world of mysteries, already become archaic, from the new alliance of empirical induction with mathematically logical thought."

That is true as far as it goes. It would be surprising in the extreme if Wolfgang Pauli were to side with the Rosicrucian alchemist Fludd against Johannes Kepler. However, it is perhaps necessary to mention that Kepler's grand astro-geometrical theories were completely wrong: as he himself showed (though he never admitted it), the elliptical paths pursued by the planets render any notion of interposing the perfect solids between them impossible; and the construct of musical ratios he describes in the fifth book of *The Harmony of the Universe* is nothing more than a glorious, exhilarating exercise in mathematical legerdemain.

Moreover, a careful study of Kepler reveals that he was a believer in the *Corpus Hermeticum* and the authenticity of Hermes Trismegistus's revelation. Frances Yates arrives at that conclusion based upon an analysis of a long passage from *The Harmony of the Universe* in which Kepler identifies the teachings of Pythagoras with the wisdom of the Hermetica, saying "either Pythagoras is Hermetizing, or Hermes is Pythagorizing" (*aut Pythagoras Hermetiset, aut Hermes Pythagoriset*). Kepler then goes on to make the conventional esoteric connection between the Hermetic-Pythagorean tradition and the book of Genesis and the Gospel of St. John. Though he was writing four years after Casaubon's dating of the *Corpus*, it is entirely possible, even likely, that Kepler still accepted the antiquity of Hermes Trismegistus. The phrase just quoted does suggest some doubt: how could Hermes have Pythagorized if he preceded Pythagoras? Of course, the phrase might also be interpreted in a purely literary way, in the sense that ideas pattern one another regardless of priority. In the epilogue to the fifth book of *The Harmony of the Universe*, Kepler mixes up homages to Copernicus and Tycho Brahe with unvarnished Neoplatonic references to Orpheus, Pythagoras, and Plato. He devotes many lines to a mystical explication of a solar hymn by Proclus, and then offers this rhapsodical justification of heliocentricity:

> [It is not] surprising if anyone who has been thoroughly warmed by taking a fairly liberal draft from that bowl of Pythagoras . . . and who has been made drowsy by the very sweet harmony of the dance of the planets begins to dream (by telling a story he may imitate Plato's Atlantis and, by dreaming, Cicero's Scipio): throughout the remaining spheres, which follow after from place to place, there have been disseminated discursive or ratiocinative faculties . . . while there dwells in the sun simple intellect, the source, whatsoever it may be, of every harmony.

Hardly the utterance of a paragon of rational materialism. Just one thing is certain: Kepler could never be accused of orthodoxy by any scientific or philosophical school, from Pythagoreanism to quantum mechanics. If we accept Wolfgang Pauli's image of him as the ultimate rationalist, we must conclude that Kepler was wasting his time chasing the chimera of the music of the spheres. For someone like Pauli, the heavens are silent as the tomb, and it was the "mathematically logical thought" fathered by Kepler himself that silenced them. Yet it was clearly Kepler's intent to use (or, where necessary, to invent) the most modern astronomical and mathematical methods to save the Pythagorean cosmic vision. He did his work only too well; after Kepler, the music of the spheres was irretrievably divorced from science, retreating forever into the shadowy recesses of esotericism. Yet Kepler was not the last great scientist to attempt to cast light into those recesses.

∾ NINE ∾

Newton and The Magic Flute

More than any other thinker of the seventeenth century, Isaac Newton has come to epitomize the impulse toward rationalism associated with the Age of Enlightenment. "Nature and Nature's laws lay hid in night: God said, Let Newton be! and all was light," was Pope's epitaph for him. For Wordsworth "Newton and his prism" were forever "voyaging through strange seas of thought, alone." In Blake's idiosyncratic pantheon, Newton was one of the chief villains; he calls the "Newtonian Phantasm" an "impossible absurdity" that foils and blunts man's imagination and leaves him in a waste land of doubt. Newton's magnificent accomplishment, the synthesis of a mathematical system which apparently rationalized the whole of creation, has intimidated all but the brightest math students ever since; we who in high school cursed Newton for having invented differential calculus were made to understand that it was child's play compared with his greatest work.

Thus Newton, a true classic, has been praised, even revered, and yet not read, which is a pity, because his was a far more interesting mind and complex personality than the conventional caricature of him as the crusty, cold-blooded rationalist leads one to believe. For example, while it is well known to every Newtonian scholar that the great man was a student of alchemy, that fact has been dropped from the standard curriculum: How can anyone as great as Newton have wasted so much of his time on something so patently wrong? In fact, he devoted a huge amount of his time to alchemical studies; his notebooks are filled with thousands of pages of astrological and alchemical writings. Of the 270 books on science in his library, there were more than a hundred on alchemy, but, aside from the works of Boyle, very few about what we would regard as pure chemistry.

Newton's amanuensis and laboratory assistant during his later years, a man named Humphrey Newton, wrote immediately after the mathematician's death that he rarely went to bed until very late, sometimes not until five or six in the morning, so deeply engrossed was he in his alchemical experiments. The fire in the laboratory was scarcely ever allowed to go out. Humphrey Newton reported that his employer would, from time to time, consult "an old mouldy book which lay in his elaboratory, I think it was titled *Agricola de Metallis*, the transmuting of metals being his chief design." Just twenty-five years later, one of Newton's first biographers, William Stukeley, had already begun the cover-up, writing that Sir Isaac had made "very important discoverys in this branch of philosophy, which had need enough of his masterly skill, to rescue it from superstition, from vanity and imposture, and from the fond inquiry of alchymy and transmutation"—which was, of course, exactly what he was himself engaged in. It was not entirely Stukeley's fault; at the time of Newton's death, in 1727, the vast stacks of his alchemical studies were labelled "not fit to be printed" and put back into their boxes, where they remained until an auction at Sotheby's in 1936.

Even Newton's masterwork, the *Principia mathematica*, which is usually regarded as the pinnacle of pure logic, contains many ambiguities. In the last decade of the seventeenth century, Newton was working on an extensive revision of the work that would incorporate a large number of commentaries having nothing to do with mathematics per se. Known as the classical scholia, these rambling historical essays, which exist in holograph manuscripts in the collection of the Royal Society of London, purported to show how Newton's discoveries were anticipated by none other than the *prisci theologi*, that line of wise men once thought to have begun with Hermes Trismegistus.

Newton's attitude toward his own work, as revealed in the classical scholia, was antithetical to some of the most basic premises of modern science. It was his firm conviction, which he intended to set forth in the *Principia*, that God revealed the eternal truths of the cosmos to a chosen handful of sages at the dawn of civilization, but that this knowledge was subsequently obscured and lost. Thus his own mathematical work, he believed, was essentially that of a modern *theologus* rediscovering the wisdom of the ancients. Scholars have always known of the existence of these scholia, but they dismissed them as mere literary excrescences, irrelevant to Newton's philosophy as propounded by the *Principia*. Modern scholars* have shown that the sheer bulk and serious tone of the scholia would tend to prove that Newton's belief in the *prisca theologia*, the revelations of antiquity, must have been deeply held. We do not know why the classical edition of the *Principia* was not published; perhaps Newton decided that it would go over the heads of his readers. In any case, there is no evidence at all that he recanted the esoteric beliefs he expressed in the scholia.

In each scholium Newton tracks down the classical adumbration of a particular point in the *Principia* and marshals it in support of his theory of gravitation. For example, as the

---

* J. E. McGuire and P. M. Rattansi, writing in the Notes and Records of the Royal Society of London, a study that serves as the basis of the present discussion.

antecedent for the propositions concerning the effects of gravity upon the planets, he invokes Lucretius: "Even the ancients were aware that all bodies which are round about the Earth, air and fire as well as the rest, have gravity towards the Earth, and that their gravity is proportional to the quantity of the matter of which they consist. Lucretius thus argues the proof of the void." Newton next quotes from Lucretius's *De rerum natura* to support his point, but the logic becomes tangled and strained beyond the breaking point, for there is in fact no demonstrable connection between the Lucretian concept of the void and Newton's own proof of gravity as a cosmic force.

In support of his premise that all matter is atomic, he establishes a philosophical family tree, in the style of Marsilio Ficino:

> That all matter consists of atoms was a very ancient opinion. This was the teaching of the multitude of philosophers who preceded Aristotle, namely Epicurus, Democritus, Ecphantus, Empedocles, Zenocrates, Heraclides, Asclepiades, Diodorus, Metrodorus of Chios, Pythagoras, and previous to these Moschus the Phoenician whom Strabo declares older than the Trojan War. For I think that same opinion obtained in that mystic philosophy which flowed down to the Greeks from Egypt and Phoenicia, since atoms are sometimes found to be designated by the mystics as monads. For the mysteries of numbers equally with the rest of hieroglyphics had regard to the mystical philosophy.

Moschus, the Phoenician sage, was identified as Moses himself by a number of seventeenth-century scholars, including Hermes Trismegistus's nemesis, Isaac Casaubon.

It is hardly surprising that historians of science have chosen to ignore such flights of mysticism: How can one seriously entertain a statement from the inventor of modern mathematics that equates numbers with hieroglyphics? Surely these strange scholia were nothing more than freaks of fancy, opportunities for the learned doctor to show off his classical erudition; a passage such as that just quoted would seem to be more at home in a quaint volume by Athanasius Kircher or some other reactionary eccentric than among the serious works of Sir Isaac Newton.

Yet the documentary evidence strongly suggests the contrary, that Newton was quite earnest in his profession of the *prisca theologia*, and in his belief that the propositions of his natural philosophy were rediscoveries of ancient wisdom. The letters and memorandums of his disciples and associates contain several explicit references to the revisions that the master intended to make for the classical edition of the *Principia*. His student David Gregory wrote a journal he called *Annotations Physical, Mathematical, and Theological*, based upon conversations with Newton at Cambridge in 1694. Here, for example, Gregory describes the changes envisioned for Book III of the *Principia*: "He will make a big change in Hypothesis III, page 402. He will show that the most ancient philosophy is in agreement with this hypothesis of his as much because the Egyptians and others taught the Copernican system, as he shows from their religion and hieroglyphics and images of the Gods, as because Plato and others—Plutarch and Galileo refer to it—observed the gravitation of all bodies towards all."

Time and again in the Royal Society scholia Newton identifies himself as a Pythagorean. The most startling scholium is one for Proposition VIII which relates directly to the great theme of the music of the spheres. Newton states that Pythagoras discovered the inverse-square relationship of harmoniously vibrating strings, and then he proceeds to assert that Pythagoras extrapolated that series of relations to the weights of the planets and their distances from the sun. This great breakthrough was expressed in esoteric terms, says Newton, and its message was obscured down through the centuries. I quote the scholium in full, not only because the reader will not readily be able to put his hands on a copy of it but also because of the insight it gives into the thought of Newton, *priscus theologus*:

By what proportion gravity decreases by receding from the Planets the ancients have not sufficiently explained. Yet they appear to have adumbrated it by the harmony of the celestial spheres, designating the Sun and the remaining six planets, Mercury, Venus, Earth, Mars, Jupiter, Saturn, by means of Apollo with the Lyre of seven strings, and measuring the intervals of the spheres by the intervals of the tones. Thus they alleged that seven tones are brought into being, which they called the harmony diapason, and that Saturn moved by the Dorian phthong [voice, or mode], that is, the heavy one, and the rest of the planets by sharper ones (as Pliny, bk. 1, ch. 22 relates, by the mind of Pythagoras) and that the Sun strikes the strings. Hence Macrobius, bk. 1, ch. 19 says: "Apollo's Lyre of seven strings provides understanding of the motions of all the celestial spheres over which nature has set the Sun as moderator." And Proclus on Plato's Timaeus, bk. 3, page 200, "The number seven they have dedicated to Apollo as to him who embraces all symphonies whatsoever, and therefore they used to call him the God the Hebdomagetes", that is the Prince of the number Seven. Likewise in Eusebius' Preparation of the Gospel, bk. 5, ch. 14, the Sun is called by the oracle of Apollo the King of the seven sounding harmony. But by this symbol they indicated that the Sun by his own force acts upon the planets in that harmonic ratio of distances by which the force of tension acts upon strings of different lengths, that is reciprocally in the duplicate ratio of the distances. For the force by which the same tension acts on the same string of different lengths is reciprocally as the square of the length of the string.

The same tension upon a string half as long acts four times as powerfully, for it generates the Octave, and the Octave is produced by a force four times as great. For if a string of given length stretched by a given weight produces a given tone, the same tension upon a string thrice as short acts nine times as much. For it produces the twelfth, and a string which stretched by a given weight produces a given tone needs to be stretched by nine times as much weight so as to produce the twelfth. And, in general terms, if two strings equal in thickness are stretched by weights appended, these strings will be in unison when the weights are reciprocally as the squares of the lengths of the strings. Now this argument is subtle, yet became known to the ancients. For Pythagoras, as Macrobius avows, stretched the intestines of sheep or the sinews of oxen by attaching various weights, and from this learned the ratio of the celestial harmony. Therefore, by means of such experiments he ascertained that the weights by which all tones on equal strings . . . were reciprocally as the squares of the lengths of the string by which the musical instrument emits the same tones. But the proportion discovered by these experiments, on the evidence of Macrobius, he applied to the heavens and consequently by comparing those weights with the weights of the Planets and the lengths of the strings with the distances of the Planets, he understood by means of the harmony of the heavens that the weights of the Planets towards the Sun were reciprocally as the squares of their distances from the Sun.

But the Philosophers loved so to mitigate their mystical discourses that in the presence of the vulgar they foolishly propounded vulgar matters for the sake of ridicule, and hid the truth beneath discourses of this kind. In this sense Pythagoras numbered his musical tones from the Earth, as though from here to the Moon were a tone, and thence to Mercury a semitone, and from thence to the rest of the Planets other musical intervals. But he taught that the sounds were emitted by the motion and attrition of the solid spheres, as though a greater sphere emitted a heavier tone as happens when iron hammers are smitten. And from this, it seems, was born the Ptolemaic system of solid orbs, when meanwhile Pythagoras beneath parables of this sort was hiding his own system and the true harmony of the heavens.

Pythagoras did nothing of the sort. As we have seen, according to Pythagorean tradition itself the Master erroneously taught that a simple arithmetical relationship existed between the weights of stretched strings and their tones, just as it did between the lengths of plucked strings of different lengths and their tones. The breakthrough that there exists a relationship of inverse squared ratios between the weights used to stretch strings and the tones thus produced was not made until the sixteenth century, by Vincenzo Galilei. Newton, just like the Renaissance humanists of Vincenzo's day, explained that away simply by asserting that the tradition had got it wrong. The Master, being the Master, must have had it right.

There can be no question that Newton was aware of orthodox Pythagorean music theory. In his twenty-third year he wrote a short treatise about music theory in the conventional Boethian vein, which he never published, though it was circulated among his friends at Cambridge. Yet during the same period Newton also made a number of interesting innovations in music theory that have only recently come to light. He was the first to use logarithms in musical calculations (more than a hundred years before Johann Heinrich Lambert would claim priority in doing so). In a commonplace book he kept at Cambridge, Newton expressed eighteen widely different scales, including Greek and Chinese examples, with logarithms. He also asserted an analogous relationship between the seven constituent colors of the spectrum of white light (one of his principal contributions to physical science, of course) and the tones of the just musical scale, though his theorizing on this point did not pan out.

Yet Newton was not as interested in pursuing these genuine mathematical explorations in music theory as he was in chasing after what to a modern observer can only seem to be the fantasy of an ancient philosophical pedigree for his work. It is interesting that as late as the turn of the eighteenth century, even so independent a thinker as Newton could still find it necessary, just as had the medieval encyclopaedists and the Renaissance humanists, to seek a classical justification for his theories, which were already so elegantly well supported by mathematical logic. J. E. McGuire and P. M. Rattansi suggest that it was Newton's alchemical studies that bent his thought in an esoteric direction; they find an analogy in the work of the seventeenth-century English alchemist Michael Maier, whose works Newton studied. Maier undertook to survey all of Greek mythology to show that it was a vast allegory concealing alchemical secrets. "Newton's interpretation of the 'harmony of the spheres' is analogous," they say, "in that it sees it as a symbolical representation of 'physical' secrets."

Newton also used Greek mythology as an integral part of the justification for his natural philosophy. He declares that Thales, who is usually called the first of the Attic philosophers, "taught that all things are full of Gods, understanding by Gods animate bodies. He held the sun and the Planets for Gods. And in the same sense Pythagoras, on account of its immense force of attraction, said that the sun was the prison of Zeus, that is, a body possessed of the greatest circuits. And to the mystical philosophers Pan was the supreme divinity inspiring this world with harmonic ratio like a musical instrument and handling it with modulation, according to that saying of Orpheus 'striking the harmony of the world in playful song.' " Once again, Newton has invented a scientific (i.e., Newtonian) interpretation of the ambiguous utterances of the earliest, semi-legendary philosophers. Then he goes on to articulate what is really nothing less than a summary, concise and faithful, however much it is "Newtonized," of the great theme of the musical universe: "The soul of the world, which propels into movement this body of the universe visible to us, being constructed of ratios which created from themselves a musical concord, must of necessity produce musical sounds from the movement which it provides by its proper impulse, having found the origin of them in the craftsmanship of its own composition." Using language almost identical to that used by Pythagoras and Plato more than two thousand years before him, Newton saw in the perfect order of music the most apt analogue of the orderly cosmos. Yet however earnest were Newton's beliefs in these sentiments, the classical scholia were never published in their intended place, side by side with the propositions of the *Principia*. While Newton had actually crossed over that line straddled by Kepler before him, perceptible only in retrospect, which divides the age of classical thought and the modern era, the point at which natural philosophy began to be governed by logic rather than belief, he still yearned to maintain a spiritual connection with the seers of antiquity. Yet however much he may have seen himself as the last of the *prisci theologi*, he has come to occupy the first position among modern intellectuals.

After Newton, the great theme vanishes from legitimate science, and in the arts it would soon be swamped by Romanticism, as the heliocentricity of the Enlightenment was replaced by defiant anthropocentricity. Yet there was one last, lovely flowering of the great theme in the eighteenth century, in the artistic expressions of Freemasonry. The Masonic brotherhood, like many esoteric societies, claims to be as old as civilization itself, tracing its foundation to the building of the great Temple of Solomon; in fact, the Masonic movement seems to have evolved out of the artisans' guilds of the early Renaissance, emerging in its modern form in sixteenth-century Scotland.

Freemasonry, known to its adepts as the Craft, came into existence at least in part as a response to the perceived intellectual tyranny of the church. Its opponents accused it of being an alternative religion, and despite all the pains that its founders took to prevent that from being the case, to a certain extent it was true, just as it was of Renaissance Hermetism. Indeed the two were inseparably intertwined. It is not clear exactly how much the Masons borrowed from the Hermetic tradition, but it is certainly true that both flowed from the same Neoplatonic source, more or less at the same time and in the same places. In his recent study of *The Art and Architecture of Freemasonry*, James Stevens Curl gives this synopsis of the core beliefs of the Masons, which also serves tolerably well as a summary of the Hermetic tradition: "Freemasonry taught that beyond a gloomy and materialistic world lay a new light-filled place towards which Mankind should strive, and that it was imperative that all men should seek to build a Temple of Humanity in which all valuable knowledge would be enshrined, and where the lost past would be remembered." One feels, for example, that Giordano Bruno and Tommaso Campanella would have felt quite at home in such a place.

One of the most basic legends of the Craft is that of the Two Pillars, which is integrally linked with the tradition of the great theme. The tale of the Two Pillars occurs in the Apocrypha, although it seems to have had a Babylonian origin. Many versions exist. Essentially it relates that the most fundamental knowledge of the cosmos, the *prisca theologia*, which was revealed by God to the first men, was inscribed upon two pillars, one of marble and the other of brick. Astronomical discoveries were on one pillar, and the secrets of music were carved upon the other by Jubal, the father of music and musicians according to Genesis. According to Masonic lore, after the Great Flood the Two Pillars were discovered, one of them by Pythagoras and the other by Hermes Trismegistus, who imbibed this secret knowledge and passed it on through their philosophies.

Though the origins of Freemasonry are unclear, for our purposes it is sufficient to postulate that it may have emerged in Scotland during the period of the Reformation. By the eighteenth century the Craft was well established throughout Europe. Its members came from every stratum of society, including the very highest: for example, the husband of the Archduchess Maria Theresa, Francis Stephen, Duke of Lorraine, was an initiate. The Masons identified themselves explicitly with the concept of Enlightenment, in the rather literal sense of "Fiat lux," Pope's motto for Newton. Newton, in fact, was a heroic figure to the Masons. One of the most remarkable designs of Masonic architecture was the cenotaph for Newton designed in 1784 by Étienne-Louis Boullée. This monument took the shape of an enormous hollow sphere, a shape that had the same important symbolic connotations for the Masons that it did for the Pythagoreans, within which was suspended a blazing globe suggestive of an armillary sphere. Boullée composed this text for the cenotaph: "Esprit sublime! Génie vaste, et profond! Être Divin! Newton!" (Sublime spirit! Genius vast and profound! Divine Being! Newton!) The monument was never built, and it probably would have been impossible to construct, from an engineering point of view. Boullée's monument is perhaps unwittingly appropriate: while such an apotheosis of Newton might on the face of it seem to arise from a misinterpretation of the creator of the mechanical universe, by linking him with the mystical tradition to which Freemasonry was heir Boullée tapped into that other, alchemical side of Newton, the one revealed by the classical scholia.

Undoubtedly the greatest artistic achievements in which Freemasonry played a part were the Masonic compositions of Mozart, who was initiated into the Craft in Vienna in 1784. Mozart loved Freemasonry, finding in it a camaraderie that appealed to his gregarious nature and a solace to palliate his melancholy moods. He wrote some dozen works with specifically Masonic content, including several cantatas and the exquisite *Maurerische Trauermusik* (Masonic Funeral Music). The most famous of Mozart's Masonic works is *Die Zauberflöte*. *Die Zauberflöte* is usually characterized as a sublime paean to the Enlightenment, all sweetness and light—and it is. Yet it is also an earnest attempt by two loyal initiates of the Masonic brotherhood, Mozart and Emanuel Schikaneder, the theatrical producer who wrote the libretto, to defend Freemasonry against critics in Austria who were calling for it to be proscribed.

The popularity of the Craft was soaring throughout the Empire, and the freethinking attitudes it inculcated, particularly with regard to the Rights of Man, were not welcomed by a

Étienne-Louis Boullée's *Cenotaph for Newton*

monarchy that perceived itself to be in danger. It was well known that the revolutionary leaders who had triumphed in America and those who were about to do so in France were Masons. So Emperor Joseph II ordered a reduction in the number of lodges in Vienna in 1785, the same year that Leopold Mozart, Wolfgang's father, and his friend Franz Josef Haydn were initiated into the brotherhood. The Emperor also required the lodges to provide the government with lists of the names of their members—inimical to a secret society, but invaluable to modern scholars.

Furthermore, the institution of Freemasonry had been denounced by a papal bull, although it seems never to have been enforced in the Holy Roman Empire. After the death of the emperor in 1790, the very survival of Freemasonry in Austria was in doubt. The French Revolution was beginning to have profound repercussions throughout Europe, and Marie Antoinette, formerly Archduchess of Austria, was warning her brother, the new Emperor Leopold II, that he should beware of the Masons, who she felt were behind the overthrow of her husband's monarchy. Such was the atmosphere in Vienna when Mozart and Schikaneder began their collaboration, which premiered at the latter's Auf der Wieden Theatre on September 30, 1791, less than three months before Mozart's death. The two men were determined to present the Craft, which for them was a powerful force for good, in the most favorable possible light.

Die Zauberflöte was not the first Masonic opera; Jean-Philippe Rameau's *Zoroastre* of 1749 and Johann Gottlieb Naumann's *Osiride* of 1781, to name just two, were based on Hermetical themes, though they seemed not to have influenced Mozart, a composer never in need of inspiration. Mozart's first musical brush with Freemasonry came in 1772, when he wrote two choruses for the premiere of *Thamos, König in Ägypten*, a play by the Masonic dramatist Tobias Philipp

Freiherr von Gebler. Six years later, he composed additional incidental music for a new staging of the play produced by Schikaneder.

Although the incidental music for *Thamos* was his first specifically Masonic work, in 1772, when the composer was sixteen years old, he composed a dramatic serenade for the installation of a new archbishop in Salzburg, to a libretto by Pietro Metastasio based upon *Scipio's Dream. Il sogno di Scipione* is a charming little piece and deserves to be better known, but its text is a feeble presentation of the great theme. Cicero's original fable was not rich in dramatic possibilities to begin with; but Metastasio (who was also the author of the noble libretto of *La clemenza di Tito*, which Mozart composed simultaneously with *Die Zauberflöte*) had only the sketchiest understanding of the principles underlying it.

Cicero's story, slim as it is, has been abandoned for an allegorical tableau involving Scipio the Younger and Fortuna and Costanza (Fortune and Constancy to Duty). Scipio has fallen asleep in the palace of the barbarian king (here, strangely, called Massanissa rather than Cicero's Manissa). In a dream, Fortuna and Costanza appear to him and tell him that he must choose between them. His adoptive grandfather appears, urging him to be virtuous. His father also makes an appearance, to tell him how inconsequential are the concerns of mankind. In the end, of course, Scipio chooses Costanza.

Metastasio's description of the musical universe occurs after Scipio awakes and asks where he is:

SCIPIO
*Then where am I? This is certainly
not Massanissa's palace,
where a short while ago
I surrendered my eyes to sleep.*

COSTANZA

No. Africa is very far away
from us. You are in
the immense temple of heaven.

FORTUNA

Do you not recognize it
from so many brilliant stars that shine
around you, from the unwonted harmony
of the moving spheres you hear,
from the great globe
of shining sapphire you see
that bears them off into orbit?

At this point, Scipio asks the goddesses who produces this "so harmonious a concord of sound." Here Metastasio's lack of understanding of Pythagorean principles becomes painfully obvious. Cicero's fable had a somewhat mechanical quality, but Metastasio makes a sad jumble of it:

COSTANZA

That same unequal proportion
which exists between them
in motion and size. In their courses
they come into collision: each gives forth
a different sound from the next;
and from all a harmonious sound is formed.
The strings of a lyre are likewise different,
yet the ear and hand temper
treble and bass in such a way that,
when struck, they produce sweet harmony.

This marvelous ensemble,
this mysterious ratio,
that unites the dissimilar,
is called proportion, the order
and universal principle of all creation.

> *That is what lay hidden,*
> *an arcane ray of higher learning,*
> *within the numbers of the sage of Samos.*

Although Scipio has asked "chi," *who* produces the musical concord of heaven, Costanza tells him it is "the same" *proporzionata inegualglianza*, although it is the first time that that mysterious quantity has been mentioned. The goddess's reference to the "arcane ray of higher learning" concealed in Pythagoras's numbers demonstrates that Metastasio was well aware of the Hermetic tradition, even if he was not himself initiated into its mysteries. Finally, Scipio asks the familiar question: "But why does so glorious a harmony not reach us? / Why is it not heard by those living in earthly abodes?" Costanza replies:

> *It too far exceeds the perception of your senses.*
> *The eye that turns upon the sun*
> *cannot see the sun at which it gazes,*
> *dazzled by that same*
> *excess of splendor.*
>
> *He who lives on the banks*
> *of the tumbling waters of the Nile*
> *does not notice the noise*
> *of the raging torrent.*

To turn from *Il sogno di Scipione* to *Die Zauberflöte* is to progress from youthful superficiality to the profound expression of a master in full possession of his genius. Mozart and Schikaneder's story is loosely based upon *Sethos*, an allegorical novel by Abbé Jean Terrasson, a Mason, about an ancient Egyptian prince, the eponymous Sethos, who undergoes a number of initiatory trials and travels throughout the world in search of spiritual enlightenment. *Sethos* was not only widely read by Masons throughout Europe but also sometimes even cited by French historians as an authority on ancient Egyptian religion.

The character of Tamino in *Zauberflöte* is undoubtedly Sethos himself, and the ordeals Tamino undergoes in act 2, scene 11 are based upon Masonic initiation rites. The character of Sarastro, the High Priest of the Temple, is plainly modelled upon Zoroaster, one of the greatest of the *prisci theologi*.

The Masonic nature of the opera is evident from the opening chords of the overture, which is in the key of E-flat—a "Masonic" key signature, as it is denoted by three flats and thus represents the Triad, which had the same transcendental significance for Freemasons in the eighteenth century that it had had for the Pythagoreans. The stately opening phrase—a chord, followed by two more chords, each preceded by anacruses (introductory upbeats)—creates a five-part rhythm, thus: X-xX-xX. Five combines the Dyad and the Triad; moreover, according to James Stevens Curl, in Masonic lore five "represents the flaming star of the female Order, or light itself," which gives the opera a perfect symmetry, for the finale expresses the ultimate triumph of light over darkness. Curl's analysis of the overture continues:

> The following Adagio is a conventional representation of the Kingdom of the Night, or Darkness and Chaos. The opening of the fugue is *Ordo ab Chao*, the Kingdom of Light, with a rhythm suggesting the blows or tapping of mallets, that is, Masonic work. The fugue breaks off for the thrice-three chords of its Master's Degree with dotted rhythms . . . before a darker section of self-examination in the minor key with many chromaticisms (suggesting a journey in sound) leads to the major closing bars, representing the victory of the sun.

Masonic themes are abundant throughout *Die Zauberflöte*, particularly numerical symbolism. The number three, one of the holiest numbers since the time of Pythagoras, is everywhere in the work: besides the three knocks of the overture, there are the

Karl Friedrich Schinkel's design for act 2, scene 7, of Mozart's
Die Zauberflöte

Three Ladies, the Three Boys, and many instances of triplet rhythms.

One of the mysteries of the opera is why the Queen of the Night, who was originally a benign character, was altered into the shrilly vengeful lunatic of the final version. One likely explanation is that Mozart and Schikaneder used her to symbolize the church. Karl Friedrich Schinkel's famous design for act 1, scene 6, depicting the Queen ascending into the vault of heaven on the crescent moon, surrounded by orderly ranks of stars, is suggestive of the Virgin Mary in her apotheosis as the Morning Star. Thus the Queen's implacable hatred of Sarastro, the High Priest, is allegorical of the church's opposition to Freemasonry; when Tamino finally realizes that the brotherhood of priests (i.e., the Masons) is a force for the three virtues of Strength, Beauty, and Wisdom, as the Three Boys have promised, it suggests that Freemasonry will overcome its adversaries. Mozart and Schikaneder's hopeful prediction did not come to pass; Freemasonry was outlawed in Austria in 1795, and not made legal again until 1918.

Several books have been written to unravel all the Masonic symbols hidden in the text and music of Mozart's opera (Jacques Chailley's *The Magic Flute, Masonic Opera* is especially thorough), but it is not necessary to dig deep into the recesses of the opera to find its connections with ancient concepts about the musical universe. At moments the precepts of the great theme shine forth with sparkling clarity. When Tamino plays the magic flute in act 1, scene 12, enticing the wild beasts to come forth to hear his song, he is clearly being identified with Orpheus. Later, in act 2, scene 11, the Two Armed Men sing:

> *He who shall tread this path so full of trial*
> *Fire, water, air, and earth shall undefile.*
> *And when he shall have conquered death's fear*
> *Then shall he rise to heaven's sphere.*
> *Illumined shall he be, and consecrate.*

Die Zauberflöte probably represents Mozart's own philosophy more faithfully than does any of his other operas. Writing for a production at a popular theatre in the suburbs, in collaboration with a friend and fellow-Mason, the composer did not have to worry about satisfying the vanity or furthering the political agenda of a noble patron. Nowhere else in his oeuvre do we feel so close to Mozart as in the finale of the opera, when Sarastro, standing before the Temple of the Sun, sings:

> *The sun's golden splendor now sunders the night*
> *And shatters the power of the evil one's might.*

≈ TEN ≈

The Romantic Anomaly

If this book until now has been weighted more heavily with examples of how musical values have infused the natural philosophy of Western civilization than with descriptions of the celestial theme in music, that is because the latter is so pervasive: a history comprising all the instances of the cosmic theme in music would itself almost constitute a history of music. To say, for example, that the late works of J. S. Bach are characterized by mathematical purity and cosmic scope is to state the obvious. Yet it is certainly true; in some ways Bach's works are better expositions of the great theme of celestial harmony than is, say, the explicit statement of it in *Die Zauberflöte*.

Leibniz, Bach's contemporary, formulated a definition of music that seems to have been tailor-made for Bach: "Music is the hidden arithmetical exercise of a soul unconscious that it is calculating." Bach was a learned man, and most of the sources of the great theme that I have cited were familiar to him. The twin qualities in Bach's compositions of numerical perfection and profound spirituality are remarked upon in almost every one of the hundreds of books about his works; in one study, *Bach and the Dance of God*, Wilfrid Mellers explicitly links the composer with the great theme. In this passage, Mellers begins by continuing the quotation from Leibniz:

> "If therefore the soul does not notice that it calculates, it yet senses the effect of its unconscious reckoning, be this as joy over harmony or oppression over discord." This goes to the heart of the matter. Bach's musical mathematics in his last works may be seen as exact as an exercise in the dialectical logic of Leibniz or Spinoza, but it is "truer" in that its intellectual rigor encompasses the total range of human experience.... The theologians, philosophers, alchemists, and music theorists whom Bach read encouraged an equation between mysticism, magic, and number, absorbed from Greek and Oriental sources, from scholastic philosophy and, in pseudo-scientific form, from the metaphysicians of the then present. Such concepts exerted an increasing influence on Bach. As rational Enlightenment encroached, Bach ballasted his faith with hermetic truths that could be demonstrated, in terms of music, with an exactitude that leaves verbal language helpless; what results from Bach's *demonstratio* of the transcendental unity of number is liberation and joy.

In Mellers's view, Bach's importance in world civilization has no limit: his music epitomizes the entire Judeo-Christian tradition, with Zen Buddhism, the pancultural archetypes of C. G. Jung, and the Eternal Return of Mircea Eliade added for good measure. While the universality of Bach's music is beyond question, by the time we reach the Baroque era the connections between mathematics and musical composition are so well established that we must take care not to miss the trees for the forest. Bach's music may be thought of as existing in the world of eternal forms, but we should avoid the mistake of thinking that he himself was eternal: he was neither a throwback to the medieval era nor a visionary in advance of his time, both descriptions sometimes applied to him, but rather a flesh-and-blood man with his feet firmly planted in the earth of the first half of the eighteenth century.

The dialectical difficulty is especially acute when we approach Bach's great late works, created for performance in the church. Any composition that has as its primary and immediate purpose the glorification of God must necessarily be concerned with the relationship between man and the cosmos, that is, the great theme rendered as Christian dogma. Yet that is not the eternal quality in music that concerns us here. We are rather seeking to elucidate the eternal aspect inherent in musical form. The absolute perfection of a harmonious chord is as hard to describe as the perfection of a circle or the symmetrical clarity of a mathematical theorem. The perfection of the thing is immanent in it; any attempt to elucidate, to put your finger on it, will cause it to evanesce like a puff of smoke.

Arthur Symons expressed the dilemma concisely: "The reason why music is much more difficult to write about than any other art is because it is the one absolutely disembodied art when it is heard, and no more than a proposition of Euclid when it is written." Although Symons's comment was about music as a subject for the essayist, it applies equally well to any expression on the subject. As a rule of thumb, the closest one may come toward expressing the effect of a profoundly moving musical experience is the sort of comment that most embarrasses us: "Wasn't that lovely!" There is very little in the annals of the relatively young art of music criticism that would tend to negate this point of view.

An overtly spiritual work such as Bach's *St. Matthew Passion* is, obviously, motivated by a sense of the divine. Yet how do we know that? Is it because our experience of, for example, the aria "Erbarme dich, mein Gott" is deeply affective, that it speaks to our peculiarly modern sense of despair as aptly as though it were composed yesterday? Does that not evince the presence of something eternal? While such a response may be valid, it is based upon superficial and even frivolous connections that have little to do with Bach's music. In the first place, to the extent that our reaction is based upon the aria's noble text, we are moved by something that is close to being extraneous to the musical experience, rather as if one admired a Nativity by Raphael because it presents such a sweet family scene (though that is, in fact, one of the things that it is).

Certainly the words of the aria were deeply meaningful to Bach, just as the story of Jesus' birth was a defining constant in the worldview of Raphael. Yet to view "Erbarme dich" as nothing more than an attempt by Bach to express musically the evangelist's verses (as adapted by his librettist, Picander) is to miss the point profoundly. The *St. Matthew Passion* has vastly more to do with the composer's thoughts about the tradition of musical settings of the passion, which was a thousand years old by the time he came to compose his own version, and with the musicians at his disposal at St. Thomas Church in Leipzig on Good Friday in 1729, when the work was first performed. That is not to say that the piece is not a work of divine inspiration: if such a thing exists, this must be an instance of it. Yet we may as well seek the secret of the universe in the words of a prayer by a visionary like St. Teresa of Avila: the answer may be there, but by poring over the words of the prayer, or the notes of a Bach score, you will miss it. As the Zen archery instructor will tell you, the only way to hit the target is not to take aim. The words of the prayer or the notes of the score may be an efficacious means of getting where you want to go, but it is a transcendental process. It simply defies analysis.

We might do better, if we wish to identify the cosmic element in Bach's music, to look at those compositions that most nearly approach the state of pure music, late works such as the *Goldberg Variations*, *The Well-Tempered Clavier*, and *The Art of the Fugue*. These works were not intended for public performance, and the last was perhaps not intended to be performed at all—the musical equivalent of a thought experiment by Galileo. The apparent paradox is that while such compositions may be exactly what Symons meant by "a proposition of Euclid," they are precisely the works that are most likely to be thought of as examples of the sublime in music. On the other hand, when the late Romantic composers attempted to fabricate musical Paradises—the "Prologo in Cielo" of Boito's *Mefistofele* and the apotheosis of Marguerite in the finale of Gounod's *Faust* come to mind—the result is earthbound kitsch, however magnificent in its ability to stir the audience's senses and sensibilities.

Yet the paradox is only an apparent one, for though to someone like Arthur Symons, deeply imbibing the perfumed aestheticism of the fin de siècle, "a proposition of Euclid" was a damning phrase, implying dull dryness, in the Pythagorean tradition the more nearly an expression approaches mathematics the closer it is to eternal form, and thus to God. Bach himself would most likely have been pleased to be likened to Euclid.

Symons omitted to mention a third dimension in which a piece of music exists. Besides its ephemeral life in a given performance, a sequence of tones streaming through people's eardrums and consciousnesses and then dying away forever (or perhaps taking on a new life in the form of a recording), and its Euclidean identity in the form of musical notation on the stave, music has yet another dimension that takes shape only through the course of time: its performance history. It might have seemed as though I have backtracked by bringing in Bach at this point; after all, he had been dead for forty-one years by the time of the premiere of *Die Zauberflöte*. Yet it is my intention to use the music of Bach to define Romanticism, the great emotional outpouring that overwhelmed the expressive arts in the nineteenth century—and, while it was at it, brought about the virtual exile of the great theme of cosmic harmony.

There is no doubt that there was a sea change in European thought and taste in the last decade of the eighteenth century, a decade that encompassed the death of Mozart and the birth of Schubert, the decapitation of Louis XVI and the publication of Paine's *Age of Reason*. That sea change we call Romanticism. It is a word that we all understand, though it is notoriously vague: we would be hard put to explain what the novels of Sir Walter Scott and the symphonies of Gustav Mahler have in common, except perhaps to mention vague adjectives like "exuberant" and "expressive," yet no one would hesitate to call them both Romantic. Students of literature, music, and art history have had the perils of that sort of categorization impressed upon them, and we are now trained to look for continuities more than for change. Hence the familiar image of Bach as the pathfinder, leading the way toward Viennese Classicism, of Mozart as a precursor of the Romantics, and so forth.

Nonetheless, we know that there was a fundamental revolution in taste and artistic expression at the end of the eighteenth century, and despite all the problems attendant upon its usage, we still call that revolution the Romantic movement. Hackneyed though the phrase "sea change" may be, it is an apt one: by the time we arrive at Mendelssohn's famous rediscovery of Bach's *St. Matthew Passion* in 1829, the artistic waters are completely different from what they had been precisely one hundred years before, when the piece was first performed.

We cannot bring ourselves to forsake the belief that there really was such a thing as Romanticism. Traditionally, there have been two solutions to this problem. The first approach is to load on more and more adjectives and descriptive phrases to the definition, in order to make the word accommodate all the major figures of the period, with the result that a Romantic sounds like something out of a medieval bestiary, a fanciful monster incongruously compounded from a dozen different creatures. Yet this method would only seem to bolster the notion that the word is meaningless.

The other method has been not to define Romanticism at all but rather to use the word as a synonym for the nineteenth century, with a bit of slopping over into the late eighteenth century, so as to include Wordsworth and Coleridge, and forward into the early twentieth century to encompass the ends of

Wolf's and Mahler's careers and the beginning of Schoenberg's. That approach, in addition to doing little to encourage belief in the meaningfulness of the term, has the drawback of requiring one to lump in someone like Ferruccio Busoni, who was no Romantic.

I would propose a third approach, which might bring us closer to what we mean when we use the word: to set aside the music of the period itself and seek the temper of the age in the way it treated the music of the past.

The musical genius of an age is defined by all the compositions that are created in it. Yet that is like saying that the definition of a word consists in every instance of its being used—which is unquestionably true. Indeed, it is the only infallible definition; but like Borges's map drawn to 1:1 scale, it is all form and no function. Thus we tend to define the artistic spirit of an age by settling on a synopsis of works, a canon of the most important and influential compositions that epitomize the whole. The problem with this system is that the canon keeps changing. Many of the compositions that we now consider great and essential monuments of the epochs in which they were created were almost completely unknown in their own time and did not rise to the surface until long afterward; conversely, works that one generation considers to be towering masterpieces are consigned to the ash-heap, or at least the curio shelf, by later ones. Just sixty years ago, many works of Mozart's that we now consider to be vitally important, such as the opere serie *Idomeneo* and *La clemenza di Tito*, had not been produced since the composer's lifetime; on the other hand, sentimental pieces such as *Martha*, *Mignon*, and *Louise*, now dead as the dodo, were still holding on firmly to their positions in the core repertory. Haydn, the most influential figure in European music at the beginning of the nineteenth century, was almost completely forgotten within a generation, and was not rediscovered until nearly a hundred years later. Rossini, at the time of his death considered to be among the greatest composers who ever lived, had suffered a steep decline by the early years of this century, but in the past thirty years he has made a comeback.

This familiar phenomenon of the rising and falling of reputations amounts to a constant redefinition of what constitutes the important art of a given period. Thus, if we may apply a basic principle of the scientific method, the best means of arriving at a concise and reliable definition of the temper of the times is by taking the works of one composer as a constant and examining how they are treated over the course of time. For when we speak of "the Romantics," what we mean changes subtly over time. Like a garden or an aristocracy it cannot stay the same; after a few generations, a select group of prominent elements, the Schuberts and Chopins, will be the same, but everything else will have changed, or at least be reflected in a different light. Yet when we speak of the works of Bach, we know precisely what we are referring to. That is the Euclidean side of music, the constant of our little experiment in the history of taste. The variable, of course, is the other side of Symons's dichotomy, the fleeting phantom of the musical moment. A musical performance is ephemeral, but it does not vanish without a trace: it leaves behind programs, newspaper reviews, eyewitness accounts, and other data that the historian of taste may use to construct a curve of how musicians over the course of time believed a certain piece should sound. That curve is perhaps the best measure of the musical outlook of an age.

And Bach, being among the most versatile and flexible of composers, is particularly well suited to serve as our constant, although at the time of his death in 1750 he was by no means the giant that he is for us. He was well known in Berlin, where many of his students and his son Carl Philipp Emanuel had settled and established a serious scholarly tradition, and at Leipzig, where he had been Kapellmeister. Even so, he was remembered primarily for his virtuoso performances on the harpsichord and organ, and as a master of contrapuntal composition. In 1773 that reliable, peripatetic musical reporter, Charles Burney, remarked that "all organists now living in Germany have modelled themselves on his [Bach's] school, just as most of the pianists have modelled themselves on his son, the excellent [Carl] Philipp Emanuel Bach." Indeed, the judgment at that time was that C. P. E. Bach was a better composer than his dowdy, old-fashioned father. Yet as for the rest of Europe, the German musicologist Friedrich Blume scarcely exaggerates when he says that "for the most part there was in fact nothing of Bach's work to be forgotten because it had never really been known."

Most of his works survived to the end of the eighteenth century in manuscript copies handed down from one Bach devotee to another in a process reminiscent of the cult of the Pythagorean Brotherhood or the early Christians. It was from these groups that there first emerged the image of the composer as a semidivine hero, like a patriot or saint. The first glimpse we have of this tradition is in a curious anecdote told about two of Bach's students, J. Christian Kittel and J. Philipp Kirnberger. Each of them had an oil painting of the master hanging over his piano to serve as a kind of reward or punishment for his students. If they played well, they were permitted to gaze upon the image of Bach, but when they played badly, the teacher would lower a curtain over it.

Later in the century, conditions seemed to be ripe for a wider dissemination of Bach worship, particularly with the growth of a passionate, unquenchable taste in Germany for anything savoring of the disappearing traditions of Old Germany—which Bach ultimately came to epitomize. However, there was a widespread wave of Handelomania, to use Burney's coinage, throughout Europe in the latter part of the eighteenth century, which was sweeping all before it. In Germany, Handel's tremendous popularity and influence in London was seized upon as a proof that the English, unable to produce a great composer of their own, had had to entice a German to fill the gap. By 1802, when Bach's first biographer, a professor at the University of Göttingen named Johann Nikolaus Forkel, published his *Life, Art, and Works of J. S. Bach* (which carries the revealing subtitle, *For patriotic admirers of genuine musical art*), the cult of Bach was gaining strength. Wrote Forkel, "Be proud of him, German fatherland, but be worthy of him too. . . . His works are an invaluable national patrimony with which no other nation has anything to be compared." Such crude nationalism found many adherents for Bach, who without doubt had been a solid German from his wig to his shoe buckles.

The final stage in the enthronement of Bach was accomplished when the composer was embraced by the poet, literary lion, and reformed alchemist, Johann Wolfgang von Goethe. The closest friend of Goethe's old age was a musical academician named Carl Friedrich Zelter, who had an extensive collection of Bach manuscripts, which he had inherited from Kirnberger. (It should be remembered that at that time, many of Bach's greatest works were still unpublished and almost unknown except through handwritten parts such as those in Zelter's possession. The *St. Matthew Passion* was not published until 1830, and the Mass in B Minor in 1845. The enormously influential complete edition of his works was begun in 1850 and not completed until 1900.) Zelter took his responsibility as the keeper of Bach's flame seriously, as only a German academician can. In a letter to Goethe he wrote, "This Leipzig cantor is a divine phenomenon, clear yet inexplicable. I could cry out to him, 'Thou gavest me a task to do, and I have brought thee to light again.'"

In May 1821 a pupil of Zelter's, the twelve-year-old Felix Mendelssohn, stayed with Goethe at Weimar and played Bach for him morning and night. Then when Goethe was staying in Bad Berka, his host, the mayor of the town, played Bach's organ works for the poet as he lay in bed recuperating from an illness.

Thus it was that the most sensitive soul of the age—and its preeminent cultural figure—came to revere Bach. Goethe wrote to Zelter: "It was there in Berka, when my mind was in a state of perfect composure and free from external distractions, that I first obtained some idea of your grand master. I said to myself, it is as if the eternal harmony were conversing within itself as it may have done in the bosom of God just before the creation of the world. So likewise did it move in my inmost soul, and it seemed as if I neither possessed nor needed ears, nor any other sense—least of all the eyes."

Goethe's championing of Bach gave the ultimate cachet to the cause. All that was lacking was a great public occasion to function as the ritual anointing, and that was provided by Mendelssohn with his famous performance of the *St. Matthew Passion*. Mendelssohn intended the event to be memorable and historic, and posterity has obliged him. It took place on March 11, 1829, at the Singakademie in Berlin, Carl Friedrich Zelter's own academy. Zelter gave his permission to use the hall only reluctantly; like many of the high priests of the Bach cult, he was convinced that the *Passion* was too difficult to be performed, and in any case it would be far above the audience's head.

The prices were quite steep—the best seats were twenty silver groschen apiece—yet the house was sold out within hours. The influential *Allgemeine musikalische Zeitung* promised that the concert would "open the gates of a long-closed temple," and that the audience would find itself "not in the sphere of a festival of art but of a religious high feast." Among the celebrants at the feast that day were Heine, Hegel, and members of the Prussian royal family. Mendelssohn had no interest in realizing the musical effects that Bach might have wanted: he had 158 singers in the choir and conducted a full orchestra from the piano, apparently from memory. According to Harry Haskell's history of the Early Music movement, which in his view began with that performance, Mendelssohn cut the text by fully a third and substantially revised the score in order to make the music more palatable and "modern." Haskell writes: "The tempo and dynamic markings that Mendelssohn neatly pencilled in his conducting score indicate that he placed a premium on dramatic contrasts and highly charged emotionalism. . . . Mendelssohn was no purist; he approached Bach's music as a practical musician eager to bring it to life for his contemporaries. If that meant Romanticizing it, so be it." Mendelssohn's performance of the *Passion* was such an overwhelming success that it was repeated on two subsequent evenings, again to packed houses.

The cult of Bach was now firmly established. Soon performances of his works were heard up and down Europe, in souped-up versions that the Kapellmeister of Leipzig would scarcely have recognized. Nor was it simply the music that was celebrated: engraved portraits of the gentle, fatherly old man, fiercely German in a quaint frock coat and wig, were everywhere—a tradition that has survived to our own time in the form of the little soapstone busts that are given to pupils who successfully complete their First Book.

With Bach, the great theme, at least in the form that it had existed since classical times, comes to a halting close (or, at any rate, a pause). In the late masterpieces, the sense of the divine is far from being a theme—it is the very air that they breathe. Even to say that it is taken for granted is to overstate it, for the sheer musical facility of these compositions has finally attained the same state of theoretical perfection as the eternal numerical principles they embody.

The key phrase in the analysis from Goethe just quoted is when he says that the eternal harmony converses "within itself." Yes, the harmonies are eternal, but they are also self-sufficient: the conversation takes place within the bosom of God and does not require validation in the phenomenal world. The *Goldberg Variations* were composed to soothe the sleepless nights of a neurotic nobleman, Count Kaiserling, whose harpsichordist, the eponymous Goldberg, would sit in a compartment adjoining the count's bedchamber and play the variations when his patron was afflicted with insomnia. The score of *The Art of the Fugue* does not even specify which instruments it ought to be played upon. It is arguable that Bach's intentions are as well served when the work sits upon a shelf, unread, as when it is performed in some arrangement or other, as a sort of anticipation of the conceptual compositions of John Cage.

Yet by 1829 we are in a different world. It is not just that Bach himself has been apotheosized; the musicians who performed his works were perceived to have some of the divine composer's glory rubbed off on them. One can imagine Felix Mendelssohn, just turned twenty, presiding over his little army of virtuoso musicians and choristers at the Singakademie, astounding the assembled Berliners with his prodigious feat of musicianship and memory. The audience at that performance of the *St. Matthew Passion* was receiving an image of the music, and just as was the case with the portraits of Bach that hung over the pianos of Kittel and Kirnberger, it was the image of man.

If our use of Bach as a mirror to cultural history can teach us one thing, it is that in the nineteenth century the scale and emphasis shifted from the cosmic to the human. This focus on the human in the Romantic age has two principal emphases: a concentration on the performer and composer as personality, and in the scope of the music itself. The former was the guiding principle of the age; every field had its great men, its Napoleons and Nelsons, Byrons and Goethes, who were endowed with preternatural inspiration and thus the object of worshipful acclaim. Not only was Bach more famous a hundred years after his death than he ever was during his life; it is probable that his posthumous fame was greater than that experienced by *any* musician before him during his own lifetime. The glory of even the most successful composer before the nineteenth century never reached the level of hero-worship lavished upon Bach and his companions in the pantheon, particularly Handel, in the early Romantic period.

If anything, the adulation heaped upon Handel exceeded that of Bach; for in addition to the Germans the English claimed him as an icon of national pride—a formidable nationalistic combination. Handel's operas, which had been his most popular works in London during his lifetime, were entirely displaced by the sacred oratorios, intensifying the nimbus of divinity that surrounded him. In some cases it is literally an apotheosis, as in this panegyric by Leigh Hunt: "Handel was the Jupiter of music; . . . his hallelujahs opened the heavens. He utters the word 'Wonderful,' as if all their trumpets spoke together. And then, when he comes to earth, to make love amidst the nymphs and shepherds (for the beauties of all religions find room in his breast), his strains drop milk and honey, and his love is the youthfulness of the Golden Age."

Hunt might as well have said that Handel's hallelujahs *were* the heavens: whereas it had been a basic assumption of Western culture since the time of Pythagoras that music was a reflection of the heavens, for the Romantics the heavens were patterned after the choruses of Handel's *Messiah*, the piece Hunt is thinking of here. Hunt's method of making Handel a pagan god presiding over a Golden Age, and thus skirting impiety by invoking classical literature, is familiar to us; it is borrowed straight from the early Christian apologists and Renaissance humanists. (Hunt's attribution to the composer of his own penchant for exotic philosophies might have come as a surprise to Handel, who appears not to have had a strong interest in world religions.)

Performances of *Messiah* in the nineteenth century became the aesthetic equivalents of an Orphic rite, with the composer himself venerated in the place of the Demiurge. The year 1856 saw the beginning of a series of monster performances of

Handel's oratorios in the Crystal Palace, an immense structure of glass and iron that had been erected outside London for the first International Exhibition in 1851. The apogee was reached in 1883, when *Messiah* was performed by an orchestra numbering five hundred and a chorus of four thousand; in attendance was an audience of 87,769. These concerts of Handel's oratorios at the Crystal Palace continued periodically until 1926.

The same process of the deification of the artist was taking place in the plastic arts; we now think of artists such as Leonardo da Vinci and Michelangelo as being prodigiously gifted personalities, larger than life; yet although they were unquestionably famous in their lifetimes, their fame was nothing like the sort of idolatry that nineteenth-century critics such as John Addington Symonds visited upon them. We may even assert that a figure as universally loved down through the ages as Shakespeare was never revered more profoundly than he was in the Romantic era, when Emerson did not scruple to put him into the same verse with Jesus:

> *I am the owner of the sphere,*
> *Of the seven stars and the solar year,*
> *Of Caesar's hand, and Plato's brain,*
> *Of Lord Christ's heart, and Shakespeare's strain.*

From the deification of the great artists of the past, the next step was to find divine genius among the living, and that is what happened in the late Romantic age. For the first time the musician was conceived of as an artist, a term that was beginning to take on a deeper meaning. As Alfred Einstein observes, Bach and Mozart never considered themselves artists; they were very much functional and well-integrated members of their societies, achieving modest worldly success by virtue of their abilities. Yet in the emerging Romantic art cult, the composer was thought of as a solitary individual, ruled by wayward passions, in fundamental opposition to a philistine society. The artist existed on a more exalted plane than the workaday world of his audience; he breathed a rarefied ether and consorted with those sublime Intelligences that had once been charged with turning the crystal spheres.

By no means were all nineteenth-century composers regarded as divinities; nonetheless, the concept was commonplace among the emerging class of professional critics and aesthetes. If no one worshipped Camille Saint-Saëns, to choose one of the era's medium-bright luminaries, Saint-Saëns might have described himself as a worshipper of the divine geniuses (especially those of previous eras). The most worshipped artist of them all, Richard Wagner, put the new anthropocentricity boldly at the heart of his musical philosophy with his famous dictum "As man is to nature, so art is to man." Later in his treatise *Das Kunstwerk der Zukunft* (The Art Work of the Future), he sets the concept down more explicitly: "The *true* aim of art is accordingly *all-embracing*; everyone animated by the true artistic impulse seeks to attain, through the full development of his particular capacity, not the glorification of *this particular capacity*, but the glorification *in art of mankind in general.*" Thus, for all his high-flown, sentimental talk about the spiritual profundity of Christianity, Wagner is more interested in the earthly power of religion than in its cosmic implications. Jesus was for him more significantly a Nietzschean hero than an expression of the infinite. Elsewhere in *The Art Work of the Future* Wagner weaves an extended metaphor comparing music to "the boundless sea of Christian longing," with Beethoven cast in the role of Columbus: "So through the hero who sailed the wide shoreless sea of absolute music to its limits were won the new undreamed-of coasts which now no longer divide this sea from the old primevally human continents, but *connect* them for the newborn fortunate artistic humanity of the future. This hero is none other than—*Beethoven.*"

It was in the music of this hero that Romantic anthropo-centrism first glimmers forth. In Beethoven originates the paradigm of the artistic personality, difficult and deeply indi-vidualistic, as well as the first strong emphasis on the human scale in the content of the music. Beethoven's third symphony was originally to have been called "Buonaparte," but in 1804, when the composer learned that the dedicatee had crowned himself hereditary emperor, he altered the title to *Sinfonia eroica, composta per festeggiare il souvenire di un grand' uomo* (Heroic symphony, composed to celebrate the memory of a great man). Musical compositions had been dedicated to polit-ical leaders before, but the composer's motive had always been, more or less overtly, to seek patronage. In this case it was the artist who was dispensing grace to the sovereign; moreover, that grace was partly withdrawn after the great man was re-vealed to have, in the artist's perception, a base motive. (That is the usual interpretation of the incident. A great deal of discussion has been devoted to that word *souvenire*; the contro-versy centers around whether Beethoven meant that Napoleon, by his self-coronation, had ceased to be a great man, and that he was therefore only commemorating the memory of the greatness that had been, not the man him-self.)

Music was no less metaphysical after Beethoven, but the search for transcendence turned inward. Divinity was to be found in the spirit of man, not in a remote and theoretical cosmos that could only be comprehended by increasingly ab-struse mathematics. This earthward shift resulted in a para-dox: as the emphasis was transferred to the human scale, the human agent—the artist—came to be regarded as superhu-man. Beethoven provided one Romantic archetype, that of the misunderstood genius who struggles heroically against tradi-tion and its agent, society. Yet perhaps an even more influen-tial pattern was the one supplied by the life and career of

Niccolò Paganini, the violinist who defined forever the pattern of the Romantic virtuoso.

In a series of tours from 1828 to 1831, throughout Austria and Germany, and in Paris and London, the Genoa-born violinist created the greatest sensation in the history of instrumental music up to that time. The child prodigy Mozart had been one of the marvels of Europe, but his performing career came at the end of the imperial era, and so he was only known at the elite circle of the court. Paganini, a product of the age of revolution, was a self-promoted star, a public figure who held huge audiences spellbound with a dazzling display of technical virtuosity and, above all, the powerful elixir of his own personality. Where formerly considerable effort was taken to conceal the musician's artistry, Paganini thrust it into the foreground.

In the first place, his appearance was in the Byronic mode, eccentric, fearsome, even unearthly. Wrote one contemporary observer: "I can see, with the eye of memory, the whole man before me now; his gaunt, angular form; his black elf-locks falling in weird confusion over his neck and shoulders; his cadaverous face and shaggy brows; his long, hairy hands with the veins standing out like cordage." And when he played, the effect was so electrifying that, we read again and again, it exceeded the power of human speech to convey. The following newspaper account of Paganini's debut recital in Vienna, on March 29, 1828, can stand for a multitude of such reports testifying to the numinous quality that radiated from Paganini's public appearances. This one is especially interesting for the way pseudoscientific analysis alternates with Romantic hero-worship—including a gratuitous reference to the music of the spheres:

> Those who have not heard him cannot form even the slightest idea of him. To analyze his performance is a sheer impossibility—and numerous rehearsals can help but lit-

tle. When a new planet appears, travelling along an orbit of which neither the curve nor the radius can be determined, repeated observations still lead only to hypotheses. . . . When we say that the violin in his hands sounds more beautiful and moving than any human voice; when we say that his fiery soul pours a quickening glow into every heart; when we say that every singer could learn from him—we shall not have said enough to give an impression of a single feature of his performance. . . .

Paganini's compositions are of no artificial texture; they are just pure music. The intellect is never called in to arbitrate; the heart alone holds undisputed sway. In his compositions as in his playing, the vibrations of deep feeling are discernible, in which an immortal spirit soars up into the Infinite, winged with irresistible yearnings. Even his more joyous tone-paintings are scenes in which the moon is mirrored in downland pools and the heart cries for eternal day. His jests are the smiles of a sweetly-dreaming child-angel. His Adagio is the song of the spheres. He must be heard—heard. Then perhaps he will be believed.

Three years later, when Paganini played for the first time in London, he took even the phlegmatic English by storm. The audience, it was said, screamed with astonishment and delight; even the members of the orchestra were so transfixed by the novelty and perfection of his technique, and the soulfulness of his musical expression, that they failed to notice that a vandal had set fire to one of their desks until it began to blaze dangerously, and the audience frantically pointed it out to them. The English critic Henry Chorley wrote this vivid characterization of Paganini's playing, which is the reverse image of the accounts of Orpheus and Pythagoras subduing wild animals with their lyres (indeed, Paganini was nicknamed by some scribblers the "wandering Orpheus"): "Although some French author has said that *La mélancolie est toujours friande*, it certainly never was half so delicious as it appears in this strange being's performance. He literally imparts an animal sensibility to his instrument, and at moments makes it wail and moan with all the truth and expression of conscious physical suffering."

Metaphysical language abounds in every contemporary account of Paganini, but in the place of the terminology of cosmic orderliness one finds images of the demoniac. Heine described his "supernatural features" as belonging "rather to the sulphurous realm of shadows than to the sunny world of living things." The poet described Paganini's face as having a corpse-like pallor "on which trouble, genius, and hell had graved their indelible marks." Throughout his life, Paganini was plagued by rumors that he had sold his soul to the devil for his unearthly skills on the violin.

Paganini's influence was felt most forcibly upon the emerging generation of piano soloists. When the twenty-year-old Franz Liszt saw him perform in Paris in 1831, the young pianist was inspired to a state almost of madness in his attempts to achieve the same degree of transcendental technical virtuosity on his own instrument. He wrote: "What a man, what a violin, what an artist! Heavens! what sufferings, what misery, what tortures in those four strings!"

Contemporary accounts of Liszt's recitals vary only in the particulars from descriptions of Paganini's appearances: the artist's ethereal appearance, his inexpressible technical mastery, the profound spiritual depths of expression are to be found in almost every eyewitness report. A lady who saw him perform in Saint Petersburg in 1842 wrote that she and her companion were delirious with pleasure: "And no wonder. Never had we heard anything like it; never had we been face to face with such genius, with such a brilliant, demoniacal temperament, that at one moment rushed like a whirlwind, at another poured forth streams of tender beauty and grace." It is scarcely an exaggeration to say that every virtuoso instrumental soloist since the time of Paganini and Liszt has emulated their example to some extent.

In addition to originating the image of the virtuoso performer, Paganini wrought another important change in the music world (which was also emulated by Liszt and his other devotees): everywhere he played, he doubled the existing price of admission. After the downfall of the princes who had formerly supported music with their patronage, enterprising performers like Paganini made themselves a hot commodity and charged what they liked. Before Paganini made his debut in London, a bill was posted in which it was announced that boxes would be sold for ten guineas each, an unprecedented sum of money for an evening at the theatre. The ensuing uproar ultimately forced him to come down a bit, but most of his appearances were full houses, regardless of what he charged. He was the first instrumental musician to become a wealthy man by his playing.

In the Romantic era, the public became the patron. In order for young Mendelssohn to get twenty silver groschen for seats at his performances of the *St. Matthew Passion*, and Paganini his golden guineas in London, they had to offer the public something it could not find either in church or at home, where musical performances were commonplace and free. That something was, precisely, genius, a word that first came to be used in that sense during this period. Another word that might be used in this context is "greatness." I put the word in quotation marks for before the Romantic notion of the artist-as-hero, greatness belonged to heaven. God was great, not Bach or Mendelssohn, Paganini or Liszt. Of course, their audiences still believed in the greatness of God, but it was becoming more and more of a reactive belief. The revolutions sweeping Europe were dismantling the earthly part of the Great Chain of Being; and this state of affairs at least implied that the whole system might be about to tumble—which, of course, it did. While Paganini was making his triumphal tours, Darwin was sailing the Pacific aboard the HMS *Beagle*, gathering the information that would lead to his formulation of the theory of natural selection, which cast a long, chilly shadow across the notion of man's divinity. Those elephantine performances of *Messiah* at the Crystal Palace were an attempt to drown out doubt by the sheer force of numbers, to prove by the force of society's collective will that Browning had been right when he wrote that "God's in his heaven—all's right with the world."

It is easy enough to see the transformation in the Romantic era of the composer's self-image and his rank in society; one only has to compare the relatively humble positions held by Bach and Mozart with the stardom of Paganini and Liszt (and, as we shall see, a bit later on Verdi and Wagner attained the status of institutions). However, to demonstrate the new human scale in the music itself is a more subjective undertaking; nonetheless it is a quality almost universal in the music of the nineteenth century. The cult of the artistic personality permeates the literature, continually drawing the listener's attention to the hand that holds the pen. Whereas Mozart and Haydn, when they were writing in a new, experimental style, attempted to conceal the fact, to make their compositions seem more conventional than they really were, the Romantics were concerned above all with creating the impression of being bold and original.

Nowhere is the human scale more apparent than in the musical form in which the Romantics felt most at home, the song. In the vocal literature of Schubert and Schumann, the quicksilvery moods of the music reflect the atmosphere of the words, typically a deeply personal, even narcissistic first-person account of a broken heart or a dead lover, taken from the works of an arch-Romantic poet such as Heine. In Schubert's song cycle *Die schöne Müllerin*, for example, the listener is meant to associate the whimsical, depressed young miller, on some level, with Schubert himself; and however foolish the lad's romantic longings may seem to a modern person, the composer has transformed the lovesick moonings of this particular lad into the expression of a universal emotion.

It might be asked how such a composition differs from the erotic madrigals of the Renaissance period—and it would be dishonest to claim that they are entirely dissimilar; human emotions obviously have a certain universality throughout history. Yet in a Renaissance madrigal, the jubilation of a successful lover, and the repining of a dejected lover, are generalized, addressed to or describing the coldness of "my lady"—any lady. Unlike the case of *Die schöne Müllerin*, there is no sense that this is a particular young man, to some extent identified with the composer and the singer; the madrigal is the voice of Everyman.

In the instrumental literature, the music aims to capture spontaneously the emotions of the composer's wandering fancy. Schubert's impromptus and *Moments musicaux*, and Schumann's *Fantasiestücke*, among the earliest examples of the type, stand in diametrical opposition to the keyboard works of Bach and Domenico Scarlatti, which have an inevitable, symmetrical perfection in their form. By contrast, the Romantic pianist attempts to create the illusion that he is making it all up as he goes along, as though he is whispering a confidence in our ear. The impromptus of Schubert and Chopin are the most celebrated piano compositions to be called by that name, but the word might just as aptly be applied to a large chunk of the Romantic keyboard literature.

The symphony orchestra as we know it today came into existence in the Romantic era. The symphony and concerto had only just begun to take on a recognizable appearance by the latter part of the eighteenth century; but even in the hands of Mozart and Haydn, these musical forms were performed by forces that we would nowadays consider to be appropriate for chamber music, rather than a full-scale orchestra. The concerto, which places an emphasis on the individual performer pitted against the group, had emerged full-blown in Mozart's concertante works, particularly in his piano concertos. However, the struggle between soloist and orchestra is rather a collegial and good-humored one, more in the nature of a philosophical discussion, when compared with the titanic duels of the piano concertos of the Romantic era, such as those by Schumann and Tchaikovsky. The comparison is especially striking when one sets the Mozart orchestra, even for the late concertos, next to the forces massed against the soloist in, for example, the first Tchaikovsky piano concerto. There can be little doubt that in that piece the turbulent, highly charged competition between the soloist and the orchestra symbolizes the struggle of the individual against society.

In the symphonies of the Romantics, the new human scale is sometimes quite apparent, as when the composer is telling a story by using a programmatic approach, such as the musical journey of Beethoven's *Pastoral* Symphony, the forerunner of the type, or the literary excursions portrayed by Berlioz's *Harold in Italy* and Tchaikovsky's *Manfred* Symphony. Yet in most cases, the inner perspective of the musical event is submerged, sometimes very deeply. Very often the composer is self-consciously undertaking a cosmic purpose, particularly in the large-scale symphonies of the late Romantics. Yet even when Bruckner, for example, sets out to ponder the cosmos, even when the composition appears to approach a state of "pure music," the comment so often applied to his works, it is the point of view that is really under consideration. The composer pondering, not the thing pondered, is the subject.

It is another paradox that as the symphony grew more and more personal, it became ever grander and more elaborate. In the symphonies of Bruckner, which are some of the longest and most complex compositions ever written for the orchestra, we feel intuitively that what we are hearing is the supremely artistic working out of the composer's own neuroses. Apparently, in

Bruckner's case, these were manifold; among other things, he had a numerological obsession, like Pythagoras before him and Schoenberg after him. He kept careful records of how many prayers he made each day, how many times he danced with a particular girl at a ball. When he walked through one of the parks in Vienna, he counted the number of statues, and if he thought he had missed one, he would start all over. There was nothing sublime, then, about number in Bruckner's universe: it was an emblem of chaos, a fearful enemy to be conquered. The struggle for form in Bruckner's magnificently complex last symphonies is a personal and painful process, and that pain and doubt are immanent in the very stuff of the music. Yet the last symphonies of Mozart, so much more streamlined in form and succinct in expression, are as universal as Greek tragedy— precisely because they seem to have come into existence so effortlessly. When we hear the finale of the *Jupiter* Symphony, of course we are aware of the comforting presence of Mozart's personality, but the scale of the expression is indisputably cosmic, almost as though we are hearing the turning of the crystal spheres from the celestial point of view.

None of this ought to be taken to mean that I propose that Mozart is a better composer than Bruckner (even if I think that to be the case); a judgment such as that would drag in subjectivities that are better left alone. Nor am I unaware of some examples that might be produced to disprove my case: in the eighteenth century, keyboard virtuosos would sometimes improvise musical portraits of their friends, variations which were intended to convey not only the subject's personality but even his physical presence—obviously, music on the human scale. On the other hand, if it were possible for the musical historian to construct a "cosmometer," to measure objectively the exact amount of cosmic matter in musical works, it is safe to assume that no composer would rack up a higher score than Brahms. Yet not even the composers of those keyboard portraits ever considered such *jeux d'esprit* to be of any significance; they would probably be amazed to learn that posterity remembers them (which it barely does). And all of Brahms's contemporaries readily acknowledged that he was in every way an exceptional musician, unique in his talents and in his mode of expression.

The intention here is rather to draw in broad strokes the musical genius of the period, in order to show the diametric shift it traverses: from the monarchy of Louis XVI to the Paris Commune is no more a revolutionary progress, in the literal sense, than is the conceptual distance from Haydn and Mozart, composing *Hausmusik* for imperial patrons, to Wagner, reigning as a musical archpriest at his temple of art in Bayreuth. Over the course of the Romantic age, the image of the artist is transformed from that of the tormented individual, estranged from society, to that of a social institution. As the paying audience became an established force in the musical scene, the artist began to take on the trappings of temporal power.

Nowhere is that clearer than in the opera, which began to wield some weighty influence in the affairs of the world. The operas of the early Romantic period are usually motivated by the passions of illicit love and the revolutionary fervor of young heroes—the diametrical opposite of the medium's support of established authority in its formative years. Occasionally the librettos of the early Romantic period permit the heroes to engage openly in the business of toppling corrupt monarchs, as in the so-called rescue operas, of which Beethoven's *Fidelio* is the most famous example. There were also many operas, such as Donizetti's *Il Duca d'Alba* and Verdi's *Don Carlos*, that used the recently ended Spanish occupation of the Netherlands as a metaphor for the evils of kingly oppression.

At least one opera, Auber's *La Muette de Portici*, actually helped to spark a revolution. The story is set during the Spanish occupation of Naples. When a Spanish prince seduces the eponymous deaf-mute girl, her brother and his friend swear to

avenge her honor and overthrow the hated foreign oppressors.
In the second act, they sing a stirring duet:

> *It's better to die than to live in misery!*
> *Is there any danger for a slave?*
> *The yoke that chokes us must fall,*
> *And the foreigner perish under our blows!*

Ultimately the rebellion is crushed, and the despairing deaf-
mute commits suicide (making hers surely the juiciest non-
singing role in opera). The revolutionary duet provided the
impetus to launch the Belgian Revolution: when the work was
illicitly revived in 1830 at the Théâtre Royal de la Monnaie, in
Brussels, it was received with clamorous applause, and after the
performance a mob formed and stormed the courthouse. The
ensuing insurrection ended in gaining Belgium its independence from the Netherlands, which had annexed it fifteen years
before.

Yet more often the mainspring of the plot is simply to get
the girl, a task usually made hopeless by wicked princes and
devious rivals; in the end the heroine has a pathetic mad scene,
and the lovers die by their own hands, singly or together. It is a
mistake to attach too much importance to the librettos of early
Romantic operas, for most of them borrow their stories from
best-selling novels or popular poems. And it should be conceded that Mozart had already pointed in a revolutionary direction in his collaborations with Lorenzo da Ponte, particularly *Le
nozze di Figaro*, which openly sides with the servant class against
a corrupt aristocracy. In the early nineteenth century, operas
were the popular entertainment of the day, provoking precisely
the same breathless excitement among their audiences that an
Arnold Schwarzenegger blood-and-guts epic or a Danielle Steel
miniseries does today.

The cross-pollination between literary Romanticism and

the opera was profound; for proof of that we may turn to the period's most acerbic analyst of the pernicious effects of Romantic fancy, Gustave Flaubert. In *Madame Bovary*, Emma Bovary attends a performance of *Lucia di Lammermoor* at the opera house at Rouen. Emma identifies herself utterly with Lucia, and at the conclusion of the first act she experiences a profound revelation about her own love affair with Leon, which she believes is being reborn with that of Lucia and Edgardo:

> She was filling her heart with these melodious lamentations that were drawn out to the accompaniment of the double basses, like the cries of the drowning in the tumult of a tempest. She recognized all the intoxication and the anguish that had almost killed her. The voice of the prima donna seemed to her to be but echoes of her conscience, and this illusion that charmed her as some very thing of her own life. But no one on earth had loved her with such love. He had not wept like Edgar that last moonlit night when they said, "Tomorrow! Tomorrow!" The theater rang with cheers; they recommenced the entire movement; the lovers spoke of the flowers on their tomb, of vows, exile, fate, hopes, and when they uttered the final adieu, Emma gave a sharp cry that mingled with the vibrations of the last chords.

The reader, knowing the outcome of the opera, sees foreshadowed in the scene Emma's own madness and death, which will be far less glamorous than that of Lucia.

Thus, in the time-honored way of the opera, all the passionate impulses of the age were present in it in greatly amplified form. With the expansionary growth of the opera, composers felt themselves compelled to create ever grander spectacles and more thrilling emotional climaxes, reaching a pinnacle in the excesses of the grand operas of Verdi. After the auto-da-fé and ghostly intervention in *Don Carlos*, it might have been thought that the master could not outdo himself; then four years later came *Aida*, with its supremely impressive triumphal march and ballet, and the live burial of the lovers in the finale.

The operas of Wagner, set in a mythological realm of archetypes, might seem to be an exception; yet the more sweepingly grandiose they are, the more ostensibly cosmic the scale, the clearer it is that the setting of the drama is within Wagner himself. The cosmos described in the libretto of *Der Ring des Nibelungen* is exactly defined by the dimensions of the composer's brain: an overheated mélange of fervent nationalism, deeply felt interpretations of Nietzsche and Schopenhauer, and perhaps more than a dash of anti-Semitism.

It is reasonable to assert that no composer in history has ever held a more powerful position in society than did Wagner. He began his life, true to the Romantic mold, as a flaming revolutionary; his first post of any consequence, as Kapellmeister at Dresden, ended when he became embroiled in a plot against the government and had to flee the city to avoid arrest. He did not find success until the middle of his life, for his personal extravagance, and the extreme grandiosity of his operas, made him a reputation for being impossible to work with. Then in 1864 Ludwig II ascended the throne of Bavaria, and one of the first acts of the young king was to bring Wagner to his court, where the composer was given unlimited funds to produce his operas on the grand scale he had always dreamed of.

Their relationship soon became an upside-down parody of the traditional relationship between patron and artist. Whereas the letters Mozart and Haydn and other eighteenth-century composers wrote to their patrons are filled with desperate pleas, begging for just enough money to keep the bill collectors at bay, Wagner's letters have the self-confident tone of a regent chancellor advising a naïve princeling. For all that they carry flowery addresses to "My heaven-sent, dearly beloved and adored friend! My lord and most gracious King," it is plain enough that the composer held the upper hand in their dealings. At least in the beginning, no expense was spared to bring Wagner's operas gloriously to life. Ludwig's deep, neurotic attachment to Wagner eventually created a scandal, for the king's advisers came to the conclusion that the monarch's obsession with opera was jeopardizing the welfare of the kingdom.

There is no reason to believe that Wagner deliberately exploited the young monarch: the fact that he was virtually being worshipped, even by his own king, must have seemed to Wagner an eminently proper state of affairs. By the end of his life, Wagner had established himself as a despot, presiding over a little cult of artists and craftsmen devoted to the faithful production of his operas. On the eve of the first production of *Der Ring des Nibelungen* at Bayreuth, Wagner posted this notice to the company:

Final Request
to my dear fellow artists

!Clarity!
The long notes will take care of themselves; the small notes and their text are what matters.

Never address the audience directly, but always the other character; in monologues look up or down but never straight ahead.

Final Wish:
Remain loyal to me, my dear friends!

Richard Wagner

Wagner was respected and even revered, but he was not loved. That distinction was reserved for his twin genius in the opera of high Romanticism, Giuseppe Verdi. Whereas Wagner never managed to achieve an unqualified success outside his own country, Verdi in his mature career had triumph after triumph throughout Europe and beyond. He was a central, towering figure in the Risorgimento: his very name was a potent political slogan (as a fortuitous acronym for Vittorio Emanuele, Re D'Italia); "Va, pensiero," the chorus of the enslaved Hebrews from *Nabucco*, served as the unofficial anthem of the movement; and after the newly crowned king convened the first Italian parliament, Verdi reluctantly stood for office and was elected. After his death, when his body and that of his wife were conveyed to their burial place, Milan saw one of the greatest public spectacles in its history. Two hundred thousand mourners thronged the crepe-lined streets of the city. A prince of the reigning house followed the hearse on foot, while Toscanini led a choir of eight hundred in a performance of "Va, pensiero."

Yet another paradox of the new human scale of music in the Romantic era was that even as the fame and power of composers such as Wagner and Verdi grew to unprecedented heights, the music itself came to be more and more dominated by an intellectual elite centered around the concert hall and the opera house, institutions that sprang up to accommodate them. Although the Pythagoreans and their posterity were a select and esoteric group—which, it would have seemed, could hardly be outdone in terms of exclusivity—in their scheme, music per se had nonetheless been for everyone. While the adepts of the Brotherhood could claim to have a profound understanding of cosmic harmony, of which *musica instrumentalis* was only the palest reflection, the power of music itself was unquestioned, and its value was absolute. It was not necessary to "understand" an Orphic hymn or a mass by Josquin des Prez or, for that matter, the *St. Matthew Passion*, in order to benefit from them. Because of their inherent cosmic nature, their values were immanent and immediately apprehended by anyone who heard them.

Yet the rise of Romanticism brought with it the idea, which underlies much of the critical writings of Berlioz, that (in Oliver Strunk's précis) "music is not for everyone, nor everyone for music." In the nineteenth century the temples of art became public places, but the price for entering them became ever greater, and finally beyond the means of some people. Exquisite states of sensibility such as those embodied in the songs and chamber music of Schubert and Schumann (compositions that, in still another paradox, had begun as *Hausmusik* for universal consumption), profound allegorical music dramas such as those of Berlioz and Wagner, the overwhelming emotional outpouring of Mahler's symphonies—these could only be appreciated by a select few. There grew up a sort of implicit cultural Calvinism: either one was a sensitive soul or one was not.

Thus was born the musical snob. This phenomenon was the final levelling down of music to a human scale, for the value of music was defined not by the notes as they were written or played but rather by the receptivity of the person who heard. The sound world was now limited by the dimensions of the soul; and as the identity of the soul became increasingly unclear in the age of Freud and William James, finally there came about the suspicion that music might be nothing more than a transitory event in the brain, with no more inherent meaning than smelling a rose or cutting one's finger.

Yet the mystic strain was not to be so easily suppressed.

The Romantic Agenda

ELEVEN

Schoenberg and the Revival of the Great Theme

By the end of the nineteenth century, the forces of revolution and change that had spawned the Romantic movement, with its concentration on the noble and heroic propensities of the human character, were beginning to pull it to pieces. Like the metaphorical fog that falls over London in the nineteenth century in Virginia Woolf's novel *Orlando*, doubt descended upon the intellectual life of Europe. Freud followed on the heels of Darwin; and what man is and what his life means began to seem more and more like a series of accidents. If we human creatures were anything more than purposeless blobs of protoplasm, apparently, it was only in order to be held captive by our nightmares.

Thus the theme of the freedom and individuality of man came to take on a pathetic tone. Throughout the history of Western thought there had been no doubt, or scarcely any, as to man's essential divinity; even if humankind was incorrigibly fallible, there was always the potential of salvation through Christian religion or, as we have seen, by transcending the phenomenal world through mystical-Hermetic revelation. Yet by the end of the nineteenth century it seemed as though the human focus of Romanticism had become nothing more than a wearying solipsism, more likely to yield a sense of futility than anything of the glory envisioned by Beethoven and Wordsworth. In the music of the late Romantics (again excepting Brahms, though he is admittedly rather a titanic exception), there is a strong sense of the self contemplating itself—and finding the relationship to be a straitjacket.

In Mahler's late works, particularly, the melancholy of the artist is no longer offset by a life-affirming exuberance, as it had been in the music of the early Romantics. Even at their most neurotic, Schubert and Schumann had suggested that there are delights in life that somehow make all its terrors and disappointments bearable; the shadows were dappled with sunbeams. Yet by the time we come to the last movement of Mahler's Ninth Symphony, the sense of oppression is overwhelming. For all that the musical idiom has the sweeping exaltation and surface brilliance of high Romanticism, the cumulative effect is a sense of crushing chagrin, almost of calamity. By symphony's end, the audience feels as though it has just concluded an especially traumatic session with the psychiatrist: drained, nerves jangled, and yet still with no answers.

At the core of the experience, finally, there is a hollowness. One feels that it has not been any more satisfying for the composer than it has been for us—the search for answers has ended in the realization that the questions were all wrong, or, worse, that it was not permitted to ask them. The composer's yearning has expanded to the outer limits of the universe, which he finally comes to feel cannot contain all that he yearns for, because the universe is precisely himself; what he longs for most is to transcend and escape his own selfhood. Yet there is no place left to wish for, no exit possible, because his yearning, his wounded soul has absorbed every atom of what is. The artist became the victim of a cruel paradox: even as his self-conscious sensibilities grew ever more exquisite, the objective image of the universe being pictured by science left less firm ground for him to stand upon.

Then, in the first decade of this century, a revolution took place in the world of mathematics that would reverberate up and down the intellectual universe: Albert Einstein showed a way of restoring order to the cosmos. While it was necessary to chip away at the edifice of Newtonian physics in order to do so, he was able to unify time and space into a continuum and codify the laws of the universe in a concise mathematical expression. In other words, he was able to do with "real" science what the alchemists had been trying to do with "philosophical" science. At long last it seemed as though chaos might be conquerable.

The same year that Einstein published his Special Theory of Relativity, in 1906, Arnold Schoenberg composed his Kammersymphonie No. 1, op. 9, which began the process of dismantling the laws of tonality that had ruled Western music in one shape or another since its inception. With a godlike fiat, Schoenberg swept aside the rigid system of key signatures that had given harmonic coherence to music—the musical equivalent of Newtonian physics. While the increasing chromaticism of the Romantic composers, had begun to break down such distinctions, Schoenberg's move toward atonality, as it came to be known, was as profound a revolution in music as what Einstein was bringing about in mathematics. When the Kammersymphonie was first performed in Vienna, it provoked a riot.

In order to comprehend how revolutionary Schoenberg's conception of atonality was (though he hated that word, preferring "atonical" or "pantonical," the latter suggesting a merging together of all keys), it must be borne in mind that he was himself the last great Romantic composer. In the first year of the twentieth century, he composed one of the most massive pieces of music ever written, *Gurrelieder*, an oratorio setting a suite of rapturous poems by a Danish botanist named Jens Peter

Albert Einstein and Arnold Schoenberg at Carnegie Hall, April 1, 1934

Jacobsen. The piece is based upon a medieval legend about a king who falls in love with a peasant girl, who is killed by a hawk of the jealous queen.

When the work had its premiere, in 1913, it marked an unsurpassable outer limit of Romantic musical composition, both in terms of what was possible and what was meaningful. Just three years earlier, Mahler's Eighth Symphony, nicknamed "Symphony of a Thousand," had established itself as the most colossal (and colossally difficult) symphonic work in the orchestral repertory, but Schoenberg topped him. *Gurrelieder* demands five soloists, a speaker, three four-part male choruses, an eight-part mixed chorus, and a vast orchestra, including eight flutes, ten horns, and seven trombones (not to mention, among other percussion, six timpani, tam-tam, glockenspiel, xylophone, and iron chains). Given the physical dimensions of the concert stage, until the advent of electronic sound it would not be possible to make a "bigger" or more palpably affective sound than that created by *Gurrelieder*; nor did that seem like a particularly fruitful or interesting means of expanding the expressive vocabulary of music.

Based in part upon the experience of bending the huge *Gurrelieder* orchestra to his will, Schoenberg formulated the dictum, "Tonality does not serve: it must be served." By his refusal to serve, he effectively melded together melody and harmony into a single, coherent (or at least indivisible) entity. Later he would call the Kammersymphonie "the perfect amalgamation of melody with harmony, in that both of them participate equally in melting down more outlying tonal relationships to form a unity, and draw logical conclusions from the problems with which they have landed themselves." That notion of forming a unity has sometimes been likened to the Einsteinian time-space continuum. The parallels are subjective but close enough to ring true.

Schoenberg was once asked about his affinities with Einstein, and he replied, "A problematic relationship between the science of mathematics as expressed by Einstein and the science of music as developed by myself? There may be a relationship in the two fields of endeavor, but I have no idea what it is. I write music as music without any reference other than to express my feelings in tone. . . . All I want to do is express my thoughts and get the most possible content in the least possible space."

The first observation to be made about this statement is the equivalence Schoenberg seems to draw between feelings and thoughts. While the very word "feelings" is Romantic—try to imagine Bach making such a statement—by equating them with thoughts, Schoenberg suggests that everything is quantifiable and capable of being stated finitely, in mathematical terms, as it were. That is the antithesis of Romanticism, which was always aspiring to express the inexpressible: the beauty of art subsisted in the heroism and nobility of the artist's struggle. The understatement in the last sentence, "All I want to do," must be disingenuous; Einstein might as easily have said that his only interest was in solving a few problems, and that the invention of a new mathematical language was merely a convenient device for him. Schoenberg's desire to get the most content in the least space actually puts the proposition into the form of a problem of physics. Of course, that is what all composers have wanted from the first time human ancestors blew on conchs and banged two sticks together—to express themselves. But the liberation of music from its conventional formal constrictions in order to squeeze all those thoughts and feelings into the least possible space constituted nothing less than making a science of music.

Yet, paradoxically, for all his rationalist posturing Schoenberg might well be the most mystical musician since Pythagoras himself. What he shares with Einstein above all is cosmic scope, even though his method more often resembles that of an alchemist than a mathematician. The work immediately following the Kammersymphonie was the Second String Quartet, op. 10, in which he took the final, decisive step toward dissolving

the tonal foundations of Western music. While the first three
movements of the quartet have conventional key signatures, the
fourth does not: it ascends into the brave new world of pure
atonality.

Another innovation of the piece is that it sets two poems by
Stefan George, a young German poet, for soprano. That was
clearly a homage to Schoenberg's idol, Mahler, who incorpo-
rated song into so many of his symphonies. The poem of the
atonal fourth movement places Schoenberg firmly in the mystical tradition, with language that amounts to a Symbolist gloss of
Pythagoreanism:

> I feel the air of other spheres . . .
> I dissolve into tones, circling, wreathing . . .
> yielding involuntarily to the great breathing. . . .
> The earth shakes, white and soft as foam.
> I climb across huge chasms.
> I feel as if I were swimming beyond the farthest
> cloud in a sea of crystalline brilliance.
> I am only a flicker of sacred fire.
> I am only a mumbling of the sacred voice.

It is never a safe thing to ascribe to the composer the ideas
embodied in a text he has chosen to set, yet in the case of
Schoenberg, who so frequently wrote his own texts, one may do
so rather more confidently.

Another aspect of the quartet links the piece with the
esoteric tradition: it appears that Schoenberg had a secret program in mind when he composed it. Although he never revealed what the program was, there are some indications of the
general shape of this occult intention. While the George poems
are the most obvious pointers, there are certain other hints that
have been unearthed by scholars. According to Schoenberg
biographer Willi Reich, in the second movement there is a quotation from a Viennese street song, "O du Lieber Augustin,

alles ist hin!" which was a part of this secret program. One of Schoenberg's American students, Dika Newlin, said that the phrase "alles ist hin" (all is lost) was meant to be taken literally. Exactly what the composer had in mind is not likely ever to be known, but the phrase bears a striking resemblance to one of the crucial stages in alchemy.

In 1915, while he and his disciples Alban Berg and Anton Webern were immersed in the process of creating a new musical order out of the ruins of tonality, Schoenberg wrote the poem for an oratorio, *Die Jakobsleiter*. Schoenberg never completed the musical setting; in 1921 the poem itself was given a reading at the Society for Private Musical Performances, a group he founded in Vienna in 1918, immediately after Armistice. According to Berg, "The Society was founded in November 1918 for the purpose of enabling Arnold Schoenberg to carry out his plan and give artists and music lovers a real and exact knowledge of modern music."

The poem of *Die Jakobsleiter* has the outward form of a conventional sacred oratorio about the power of prayer, but its content, like that of the Second String Quartet, has a decidedly esoteric cast. The piece is divided into two parts. In the first, the archangel Gabriel exhorts the soloists and choral groups that constitute the "characters" of the oratorio to move ceaselessly, like the angels ascending and descending the ladder in Jacob's dream, as described in Genesis. Yet these celestial angels must not ask whither they are going: "Whether to right or to left, forward or backward, uphill or downhill—you must go on, without asking what lies before or behind you." Such an annihilation of the individual will is, of course, a fundamental aspect of mysticism. In the second part, the angels are the souls of the dead, who are "either accepted into higher spheres or sent back to earth in new incarnations." The conceit of dividing the unjust and just, sending the just to heaven and returning the unjust to earth, is taken directly from Plato's Myth of Er.

In the first part of the oratorio, Gabriel engages three souls who have gone astray for different reasons: a lover of beauty, a rebel against the power of God, and "one wrestling" with a thoroughly modern sense of guilt. In opposition to these three Gabriel introduces "the Chosen One," who must be Schoenberg himself: "Approach, you who . . . are a likeness, endowed with the true splendor; who resembles One far higher being, just as the distant overtone resembles the fundamental, whilst others, deeper, themselves nearly fundamentals, are farther removed from him, as a glittering rock crystal is further from a diamond than is pure carbon!"

This text explicitly sets forth the essential ingredients of the great theme: the relationship between man and god—that is, the Great Chain of Being—is likened to the series of musical overtones (and a note of modern mineralogy is added for good measure). The Chosen One, "endowed with the true splendor," in this case Arnold Schoenberg, takes his place as the last in the line that begins with Orpheus and Pythagoras, the Hermetical sages of the Renaissance and the esoteric societies of the Enlightenment. Although posterity has elevated both Berg and Webern to a status at least equal to that of their master, they always maintained a position of respectful subordination in their relations with him. Unlike Wagner, who believed that the emotional experience of hearing his compositions was a sufficient philosophical justification for his art, Schoenberg, Berg, and Webern placed the emphasis on revelation. The writings of these three men at moments imply that Schoenberg's principles of composition provide the means of divining the very meaning of life.

In the years after the Great War, they developed the twelve-tone method of composition, which for better or worse wholly transformed the musical scene of the twentieth century. The first work to exemplify the new system was the *Five Piano Pieces*, op. 23, published in 1923. The story of how the system was developed, the question of Schoenberg's priority (a composer named Josef Hauer was independently developing his own twelve-tone method at the same time), and the essentials of the system have been frequently described, so I shall offer only a brief survey.

Twelve-tone composition is the logical extension of the atonal principles that preceded it. All twelve notes of the chromatic scale being exactly equal (Hauer fancifully equates them with the twelve modes of classical Greek music), it follows that each of the twelve tones must be used equally in a given composition. To this end, Schoenberg came up with the tone row, in which all twelve tones of the chromatic scale are used, each of them only once. The twelve tones may be arranged in any sequence, with whatever rhythmic or dynamic shape the composer desires. Notes may appear simultaneously or successively, and transpositions to any octave are permitted, but the actual sequence of the row, once it has been established, must be respected. The tone row may be manipulated in three ways: by inversion (turning it upside down), in reverse or *cancrizans* (crabwise, or backwards), or inverted and reversed at the same time. While the method might seem to be restrictive, converting the art of musical composition into a sort of mathematical puzzle, in fact most of the composers who have used it seem to have been of the opinion (at least at one point or another in their careers) that, far from being restrictive, it is actually a liberating technique. A short list of those outside of Schoenberg's immediate circle who wrote compositions using his method might include Luciano Berio, Pierre Boulez, Benjamin Britten, John Cage, Hans Werner Henze, György Ligeti, Witold Lutosławski, Olivier Messiaen, Luigi Nono, Karlheinz Stockhausen, and, perhaps most important of all, Igor Stravinsky.

It is almost impossible to overestimate the profound importance that Schoenberg attached to his development of the twelve-tone system. *Doctor Faustus*, the roman à clef about him by his friend and fellow exile in California, Thomas Mann, gives some idea of Schoenberg's extraordinary solemnity and self-consciousness as a pioneering artist. Another glimpse of his towering ambition is given in the text he wrote for his alter ego in *Die Jakobsleiter*, the Chosen One. In the Chosen One's passionate monologue, he says, "I shall destroy myself in profitless battles, losing super-earthly joy and brightness and faith, love and hope. . . . I shall again have to believe that I stand alone, thrown back on myself, deserted and betrayed. . . . I shall have to say what I should never have dared to think, do what I should never have dared answer for." In an essay about the oratorio, the poet Berthold Viertel, probably expanding upon remarks made to him by the composer, explicitly sets forth the full extent of Schoenberg's intentions: "A law-maker. It is not only music that receives new laws from him. *Die Jakobsleiter* looks like the catechism of a new religion. Of Schoenberg's religion."

Schoenberg's masterpiece in the twelve-tone system was his opera *Moses und Aron*, a vastly complex score based entirely upon a single tone row. In Schoenberg's libretto, Moses is unable to deliver the law to the Israelites except through Aron, his brother and mouthpiece; but when Moses is on the summit of Mount Sinai, receiving the law from God, Aron betrays Moses and leads the Israelites to worship the idol of the Golden Calf. There can be little doubt that Moses the lawgiver, in private conversation with God, is Schoenberg himself; while the heedless, earthbound Aron is symbolic of the musical performers the composer had to rely upon to have his works brought to life. Like *Die Jakobsleiter*, *Moses und Aron* was left unfinished; although Schoenberg never stated it explicitly, it would seem that he left these two sacred works, the most ambitious projects of his mature style, incomplete not because of insufficient inspiration but on purpose. It is almost as though it would be impious to give a final and definitive form to matters so profound and spiritual; like the name of Yahweh they are ineffable. For Schoenberg, a devout Jew, such a consideration would not have been a trivial matter.

While it might seem a frivolous comparison, the parallels

between Schoenberg and Pythagoras are manifold. The composer's self-image as a lawgiver is the most obvious: we have no way of imagining what Pythagoras's personality was like, but he too must have had the unshakable sense of mission and the mistrust of outsiders that possessed Schoenberg throughout his life. The intense, fanatical loyalty of Schoenberg's students is another likeness. The composer had a penchant for confiding bursts of gnomic wisdom to the members of his circle, who would then broadcast the Master's pronouncements, typically in volumes of worshipful essays dedicated to him. That is almost precisely the method by which the philosophy of Pythagoras was transmitted.

Yet perhaps the most fundamental identity between the two men is their veneration of the numinous power of number. Schoenberg's life and works are filled with evidence of his belief in the divine power of number. The very twelveness of the twelve-tone system has magical power: the two digits added together make three, the Triad, one of the perfect numbers. The Triad, the number for things with a beginning, middle, and end, represented the union of one and two, odd and even, male and female, and the other Pythagorean dichotomies.

Moses und Aron is shot through with numerological symbolism. In the opera's first line, Moses addresses God as "Only one, infinite, thou omnipresent one." When he asks the Voice of God, emanating from the burning bush, where he will find the strength to lead the Israelites, the Voice tells him, "United with god in oneness." At the end of the first act, when Moses has gone up to Mount Sinai to achieve that oneness with God, the Israelites ask, "Where is the infinite? Where is Moses?" Thus a numerological equation has been established: Moses + God = 1 = ∞. The dynamic tension between the unlimited oneness of Moses and God and the dyad of Moses and Aron summarizes the underlying concept of the opera: the dichotomy between the sublimity and harmony of the infinite, and the discord of the finite, the earthly.

Non-German speakers often suppose that the spelling of Aron's name in the opera is taken from the German Bible; but in fact in German the name is spelled just as it is in English, Aaron. Schoenberg spelled it with just one *a* so that the number of letters in the opera's title would come to twelve, his favorite number, and above all to avoid thirteen, a number for which he had a deep loathing. When he marked the measures of his scores he would number them 12, 12A, and 14. "It's not superstition," he would say, "it is belief." He had a particularly deep fear that he would die in a year of his life that was a multiple of thirteen; on the eve of his sixty-fifth birthday he had a special horoscope drawn up. After he survived that year, he thought it was smooth sailing until he was seventy-eight; but when he turned seventy-six, a friend pointed out to him that the two digits added together made thirteen.

As is the case with Pythagoras, the death of Schoenberg is shrouded with legends. It is a fact that he died on Friday, July 13, 1951, in his seventy-sixth year, just a quarter of an hour before midnight. One account has it that his wife leaned over him and whispered, "You see, the day is almost over. All that worry for nothing." Then he died. Another account has it that he whispered a final word in parting: "Harmony." In either case, it is an eminently Pythagorean end to the modern prophet of the mystical power of mathematics in music.

Standing in stark opposition to Schoenberg is the German-born composer Paul Hindemith. Throughout the middle part of the twentieth century, he was the staunch defender of tonality—despite the fact that, like every young European composer of his era, he had been strongly influenced by Schoenberg in his early career. In the years after the Great War, Hindemith was proscribed by the Nazis after his comic opera *Neues vom Tage* (News of the Day), conducted by Otto Klemperer, which included among its characters a naked lady taking a tub bath. By the late thirties, Hindemith had rejected atonality as contrary to acoustical and psychological reality. In his monumental treatise *Unterweisung im Tonsatz* (The Craft of Musical Composition, 1937 and 1939), he declared that the tone row "destroyed the gravitational uprightness of traditional harmony," and he revised his early works to restore tonality to them.

Hindemith was the leading proponent of *Gebrauchmusik*, which held that music depended for its validity upon its ability to attract a wide audience. Simplicity and intelligibility were the hallmarks of *Gebrauchmusik*, which propounded the idealistic view that if the peoples of the world were able to listen and be moved by the same music, they would eventually abandon war and strife. It goes without saying that the adherents of *Gebrauchmusik*, which included Aaron Copland and Kurt Weill, had little use for the elitist mysticism of the twelve-tone system. Although Hindemith eclipsed Schoenberg in popularity during the years preceding the Second World War, by the time of his death in 1963 his influence had declined almost to nil, while the twelve-tone system had achieved a near total victory in contemporary concert music. The *Gebrauchmusik* dream of a popular music permeating the whole culture had come to pass, but it was happening not in concert halls but rather in the mass media, on radio and television, and on the phonograph, where Broadway, jazz, and pop music ruled. The notion of music as high art, "Art for art's sake," had reached its pinnacle in the twelve-tone compositions of Schoenberg, which seemed to most people deliberately obscure, addressed to an exclusive, highbrow cult—which, to a certain extent, they were.

Paul Hindemith's most grandiose attempt to synthesize his notions about tonality is a mystical opera based upon the life of Kepler called *Die Harmonie der Welt*, which is of course the same title as the astronomer's masterwork. The opera is now almost unknown, and its deliberately old-fashioned libretto, written by the composer in rhyming archaic German poetry, and conservative musical idiom do not augur well for its being rediscovered. Nonetheless, it is instructive for the present purpose to see how a composer poles apart from Schoenberg revived the great theme of the musical cosmos, in an utterly different musical style and for utterly different reasons. The opera is also important from a historical point of view, for it was the first opera to be written with a scientist as the central character.

The libretto of *Die Harmonie der Welt* is a series of dramatic tableaux from Kepler's life rather than an attempt at a cohesive unity. Some of the scenes, especially those dealing with the astronomer's mother, have considerable dramatic interest, while others deal at tedious length with theological disputes and internecine warfare in seventeenth-century Middle Europe. The text serves primarily as a framework for Hindemith to experiment with his ideas about tonality as an expressive technique. He rejects as unscientific and unprovable the classical notion that each of the keys creates its own distinctive mood. Instead he uses the relationships between different tonalities to symbolize the relationships between the concepts or characters associated with them. He explains this idea in his introduction to a song cycle entitled *Das Marienleben*:

> If within the field of expression narrowly circumscribed by text, story, spiritual significance, and other factors . . . I place one tone (with the tonality belonging to it) in the center of the tonal activity, then the other tonalities must take their place according to the degree of relationship to this central one. . . . I can even go further, and substitute for the equation *tonality = emotional state* a more far-reaching one, namely, *tonality = group of concepts*, so as to widen enormously the field of tonality symbolism.

In *Die Harmonie der Welt*, Hindemith has constructed his text around eight major characters, each of whom stands for a different planet in the Keplerian solar system. The associations are mostly rather obvious: Kepler himself is the earth, the Emperor is the sun, Kepler's loony mother is the moon, his wife is Venus, and so forth. Each of these characters has a distinctive tonality. The musical intervals between these tonalities reflect the distances between the characters, while the changes in their relationships with each other are shown by modulations in the corresponding tonalities, and so forth. It is a hugely complicated scheme that has been explicated by the Hindemith scholar James D'Angelo.

Hindemith obviously identified himself with Kepler in his aspect as the bringer of order, much as Schoenberg identified with Moses the lawgiver. James D'Angelo points out the curious coincidence (though Schoenberg, for one, would certainly not have used that word) that Kepler was born on December 27 (1571), while Hindemith died on December 28 (1963), and that Hindemith's birthday was November 16 (1895), while Kepler died on November 15 (1630). In the first volume of *The Craft of Musical Composition*, published twenty years before he wrote *Die Harmonie der Welt*, Hindemith takes an explicitly astronomical view of harmony—just as Kepler had taken a musical view of the cosmos.

For example, in order to explain his conception of musical intervals, Hindemith likens the tones to the solar system: "If we think of a series of tones grouped around the parent tone . . . as a planetary system, then C is the sun, surrounded by its descendant tones as the sun is surrounded by its planets. . . . The intervals correspond to the distances of the various planets from each other. In their melodic function, the two successive tones of an interval are like two planets at different points in their orbits, while the formation of a chord is like a geometric figure formed by connecting various planets at a given instant."

The finale of *Die Harmonie der Welt* is a feverish hallucination of the dying astronomer, in which all the opera's characters appear in their planetary aspect, costumed in golden robes. Amid lights arranged in the form of the constellations of the zodiac, they present a masque intended to evoke the splendor of a Renaissance court entertainment such as the *Pellegrina* intermedi or the *Ballet comique de la Reine*. Finally the chorus appears, apparelled in hazy mantles equipped with tiny lights representing the Milky Way. The final chorus is a rather naïve fusion of the idealism of *Gebrauchmusik* with the harmonic vision of Kepler:

> *Our gaze into the infinite cosmos*
> *Encircling us with rich and gentle harmonies*
> *Inclines us through vision, ardor, and faithful prayer*
> *To uplift our imperfect selves*
> *Higher than through logic and erudition;*
> *Until the Spirit of ultimate majesty*
> *Grants to our soul*
> *The grace to be merged*
> *Into the exalted harmony of the World.*

These sentiments, as noble as they may have been, must have seemed rather naïve in the Cold War years, with the very real threat of nuclear destruction hanging over the world. In an interview with *Newsweek* at the time of the premiere of *Die Harmonie*, Hindemith said, "Time will tell whether my telescope was focussed properly."

TWELVE

Into the Future

Nothing is more dangerous for a historian than to attempt to synthesize wisdom about the times in which he lives. As I have said, exactly what we are referring to when we talk about the music of a previous era is constantly undergoing tiny transformations that over time alter the landscape entirely: new pieces, even new composers, are always being discovered, and known compositions are always slipping from the penumbra of neglect into the pitchy darkness of oblivion. Emphases shift. The sublime genius of one generation is the kitsch of another, while supposedly minor composers are taken up as causes by performers and critics of a later era, who resurrect their compositions and elevate them to the pantheon.

If that holds true for the music of the past, how much more must it be the case with the music of one's own time. It is more readily apparent in the field of pop music than in classical music, but a perusal of the concert-hall programs of twenty years ago will reveal that the same forces are very much at work there. When was the last time you saw a piece of Roger Sessions being performed? Staples of programs in the fifties, his works are now virtually exotic period pieces. Hindemith himself is an even better example: while *Die Harmonie der Welt* was a major event when it premiered in 1957, today only Hindemith specialists even know of the opera's existence.

Thus it is foolhardy to make any too-sweeping assumptions about a topic as amorphous and protean as "the music of today." It would be futile to attempt to create a grand architechtonic synthesis out of the diverse strands that constitute the musical life of our times. Indeed, if any one word might be used to describe it, it would be "fragmentary." Yet the music scene of his own day might have seemed fragmentary to someone living in fin-de-siècle Vienna, a time and place that we now think of as being the very apex of High Romanticism, a moment of tremendous vitality and excitement in the history of music. Nonetheless, it really does seem a safe generalization that the present day has a far more disparate cultural scene, beset by contending forces that continually threaten to rip it apart—if they have not in fact already done so. Nowadays there is room for any musical style or conception whatsoever; if there is a trend, it is toward trendlessness. We have minimalists and neo-Romantics, serialists and microtonalists, computer music and *musica concreta*, those who seek to "fuse" classical music with jazz, with pop music, with ethnic music such as the tango—and that is hardly a complete catalogue.

The one thing all these movements have in common is that they attract modest followings, and even that may be overstating it. Some critics might try to persuade you that minimalism is the ism of the moment, based upon the success of John Adams's operas *Nixon in China* and *The Death of Klinghoffer*. Yet the audience for these works is a tiny fraction of that for the music of, say, Rossini or Liszt or Verdi in their lifetimes. To a great extent, their music was the music of their eras, widely known not only from the concert hall and opera house but also through amateur performances, piano scores, music boxes, and the rest of the apparatus for the dissemination of music in the days before the electronic media. I do not presume to know what the music of my time is, but I will venture to say that it is a gross oversimplification, if not actually wrong, simply to say that it is summarized in the operas of John Adams, which are known only to a small segment of the opera-going public, much less the great mass of people.

That is by no means a bad thing. The concept of a mass audience for music is a heritage of the early Romantic era we would do well to question. Neither Bach nor Mozart had huge publics in their lifetimes. The child prodigy Mozart was famous at courts, not in concert halls, and the sacred music of Bach was known to the churchgoers of Leipzig but scarcely beyond—yet their art seems not to have suffered as a result of their modest exposure. While a composer has no need of great fame, he does need an audience if he wants to communicate; yet there is no reason why it cannot be a small and enthusiastic one. If a microtonal composer generating sounds with computer programs can attract a roomful of people interested in what he is doing, then he and they will be doing precisely what composers and audiences are supposed to do. The fact that the great majority of people would find such an experience torture does not mean anything at all in absolute aesthetic terms.

Yet it does seem to matter, and the reason may be that ever since the Romantic era, with its emphasis on the heroic nobility of the artist, it has been expected that a composer will be famous. Until the era of Haydn and Beethoven, and the growth of the Handel and Bach cults, it did not signify much to the audience who had written a piece of music. As we have seen, however, in the nineteenth century the fame of the great composers was equal to or even excelled that of the most powerful sovereigns of the time. By the time of Stravinsky (and Picasso and Hemingway), the personality of the composer (and that of the painter and the novelist) came to be almost of paramount importance. While it is not necessary in any reasonable aesthetic analysis for a composer to be famous, because the cultural expectation for it is so high, the lack of powerful personalities in music in the late twentieth century—especially since the deaths of Leonard Bernstein and Herbert von Karajan—has been interpreted as a decline in music itself.

There are no towering figures in the sphere of concert-hall music because that is no longer the music that focusses the deepest imaginative energies of the culture. What we call classical music has become an elite and self-serving institution, overwhelmingly dedicated to the curatorial function of preserving the musical traditions of the past—undoubtedly an important service, yet anything but vital—and to a much lesser degree serving as handmaiden to the last wheezing, exhausted remnants of the avant-garde. As the avant-garde has become a more profoundly entrenched part of the establishment than any cultural phenomenon that it ever rebelled against, it has become the spoiled child of culture, demanding that its projects be supported by paternal society as a sort of conscience money, in order that the establishment may be permitted to indulge its decadent, atavistic taste for the war-horses of the Romantic repertory. Critics and musicologists devote immense amounts of energy to hand-wringing on this subject, trying to figure out who is to blame for the decline in contemporary music, and how to fix the problem. It is not, however, a problem; it is a historical fact, and nothing will ever change this state of affairs. A new symphony or opera will never again provoke a great public enthusiasm or cry of outrage, as it sometimes did in the last century and in the early years of the century now ending. The forms have been superseded; the distance between artist and audience has simply grown too great.

There are a number of reasons why this has come to pass. The principal one has already been mentioned—the ascent of the electronic media, which have elevated popular music, formerly known as folk music, to the status of the established culture. Another factor is Romanticism's most important contribution to culture, what Carl Dahlhaus has termed art religion. By the 1880s an absolute demarcation had been made between daily life and the aesthetic life of the soul, necessitating the development of all the elaborate apparatus of a full-blown religious sect. Plush, upholstered temples were constructed, where the faithful manifested their reverence by an awestruck attitude unknown to earlier generations. In the eighteenth century the lights in opera houses were bright enough to read the libretto—or the morning paper, if you preferred—and the audience was silent only as long as it was interested in what was happening on the stage. The concept of the concert hall or opera house as a sacrosanct zone, dim and numinous, where one was admonished for so much as clearing the throat, is a High Romantic, specifically Wagnerian, phenomenon.

With Schoenberg, as we have seen, art religion became an all-consuming cult. The cosmic scope of the twelve-tone school, its marriage of formal purity and deep spiritual expressiveness, seemed to those who came afterward to allow only two alternatives: to imitate it or to seem trivial by comparison. Similar dead ends were reached in the visual arts and in literature during the same years—how could Picasso and Duchamp, or Joyce and Eliot and Pound, ever be surpassed? The answer was that they could not, and after fifty years we have arrived at a similar fragmentation in those arts as well. Who could persuasively describe the Novel in English Today? And poetry, like the symphony, has become a noble fossil: the more determined the faithful are to revivify it, the more plainly is it revealed to be moribund.

The establishmentizing of the avant-garde, if I may call it that, began in the post-war years, when the third-generation descendants of Schoenberg and Joyce and Pound took refuge in universities. Vast, unreadable novels and tiny, enigmatic poems have been produced by professors for the edification of other poet-professors and professor-novelists. In the years after the death of Schoenberg in 1951, the music faculties of American universities became almost the exclusive province of the serialist school, an even more stringent form of twelve-tone composition. In serialism, as developed by Milton Babbitt and Pierre Boulez, among others, not only the pitches but all aspects of the musical notes in a composition were organized according to a strict program: the loudness, duration, and even the timbre of every note were serialized with the same rigor with which Schoenberg had organized the notes of the chromatic scale. The result was to impose so much order on the composition that the human touch of the composer was all but eliminated. Babbitt (who was trained as a mathematician) summed up the attitude of the serialists in the title of an essay he published in 1958: "Who Cares If You Listen?"

The Greek composer Iannis Xenakis, whose early works were in the twelve-tone school, has evolved a method of composition that draws on the very loftiest reaches of higher mathematics. He uses the mathematics of chance, symbolic logic, set theory, and new forms of calculus such as matrix calculus and vector calculus, manipulated by computers, to make what he calls stochastic music. (*Stochastic*, derived from the Greek meaning "to guess," is a term borrowed from probability theory.) In an influential treatise called *Formalized Music*, Xenakis reveals a cold-blooded attitude toward musical composition similar to that of Milton Babbitt. Most of the book is completely unintelligible to anyone who does not have a degree in mathematics, but the author reveals the general bent of his thoughts in his definition of music:

> But everything in pure determinism or in less pure determinism is subjected to the fundamental laws of logic, which were disentangled by mathematical thought under the title of general algebra. . . . Equivalence, implication,

and quantification are elementary relations from which all current science can be constructed. Music, then, may be defined as an organization of these elementary operations and relations between sonic entities or between functions of sonic entities.

We have come a long way from "the soul unconscious that it is calculating": the laws of logic have no force in the realm of the soul, and self-consciousness is their sine qua non.

Xenakis believes that the history of European music is the audible record of scientific and philosophical "attempts to explain the world by reason." The music of antiquity, he says, was causal and deterministic, "strongly influenced by the schools of Pythagoras and Plato," and in support he quotes the *Timaeus*: "For it is impossible for anything to come into being without cause." Then, he declares, there was a revolution. Causality underwent "a brutal and fertile transformation as a result of statistical theories in physics." In other words, science discovered that, in point of fact, the *Timaeus* was dead wrong: things generally come into existence with no causality whatsoever. Genes mutate randomly, and subatomic particles decay according to no fixed program. Ultimately, the whole notion of an orderly cosmos ruled over by a Divine Intelligence is just a sentimental delusion. Xenakis would make the mathematics of chance work on behalf of art, to give musical compositions the same inevitable and absolute correctness as a mathematical expression in physics. In this thinking, form achieves a total victory over content. It does not matter what you say, because there is nothing worth saying except what can be proved by logic, and that, by definition, is what is obvious and hence need not be said.

Given that attitude among the institutionalized composers of art music, it is scarcely surprising that the repertory of the symphony orchestras and opera companies of North America and Europe should have become ossified, relying for most of their repertory on the great flood of music from Mozart to Mahler. The finger pointers seeking to fix the blame for the "problem" always seem to end up condemning the audience: tasteless, unimaginative philistines, so goes this line of thinking, they insist upon hearing their safe and familiar favorites, and obstinately cover their ears to the music of their own times. We may set aside for the moment the question of why such a preference might not be a legitimate one; after all, aficionados of Baroque architecture and the readers of eighteenth-century novels do not come in for withering scorn, so it is not clear why the lovers of the music of those eras should be held in contempt. Yet surely it is understandable that the patrons of those institutions, who live in the real world, not in the world of Xenakis's pure logic, and who were brought up with the semi-religious belief that such music was necessary for their spiritual and intellectual growth and fulfillment, might balk at being asked to foot the bill for cold, empty music that has no intention or desire to communicate—and that furthermore clearly suggests that the audience is not quite smart enough to understand it in any case. The arrogant attitude of a Xenakis or Babbitt, abetted by critics who have every interest in demonstrating that *they*, at least, are smart enough to understand the music, has done more to reduce the mass audience for new music than any number of insufficiently deep-thinking millionaires.

Of course, there is serious music being composed now that does have the intention of communicating, and even of entertaining. The success in 1991 of John Corigliano's opera *The Ghosts of Versailles* at the Metropolitan Opera comes to mind. Yet while this complex, ambitious work elicited a highly favorable response from the audience and most critics, the comment most frequently heard was how extraordinary it was not that it had such a success but that it was done at all. It was the first new work performed at the Metropolitan in a quarter century; and while the Met is a notoriously conservative institution—its directorate considers its principal raison d'être as being to serve as curator of the repertory from Mozart to Berg—the statistical record of those institutions that do traffic in new works helps to explain the Met's conservatism. The New York City Opera, a considerably more venturesome opera house, had twenty-two premieres of new works over the same twenty-five years. Impressive, until one considers that of that number only four of the operas were repeated (with one revival each). The other eighteen pieces were given expensive full productions, made the subjects of serious discussion and analysis for a week or two, and then disappeared, never to return. The twentieth-century works most frequently revived at the New York City Opera in the past quarter century have been works of the Broadway stage, for the painfully obvious reason that they sell tickets, whereas experimental, post-modernist operas do not.

The age is forever past when a serious composer could support himself by writing lively, enjoyable music for a royal supper or boating excursion, or a mass for the chapel royal. To be a composer today entails a far greater degree of self-conscious seriousness than was ever the case before this century—again, because of the deep division between high and low culture that crept in with the Romantic era. While phrased with characteristic hyperbole, Leonard Bernstein's famous statement that Lennon and McCartney were the greatest songwriters since Schubert is right on the money. Nonetheless, when a contemporary composer does find within himself both the deep desire to communicate and the technical facility and authorial conviction to do so—as was the case with, say, Benjamin Britten and Olivier Messiaen—then the audience will follow him, however difficult the journey proves to be. I do not suggest that either of those composers is as popular as Mozart or Verdi; but there are people who care passionately about their music, and they will not permit it to die. On the other hand, while I do not wish to seem to be carrying a grudge against either Babbitt or Xenakis, it is difficult to imagine that there would be a great outcry if their music ceased to be performed, except perhaps from a handful of critics. I do not say that their music is not worth performing, only that very few people care whether it is performed or not.

I have asserted that the musical scene of the present day is a pluralistic tent that can accommodate any musical style or conception. Yet if we look beneath the surface (sometimes far beneath, sometimes not), we shall find that most of these diverse strains are vestigial remains of musical schools and ideas from the past. For example, the mystic strain, the occult tradition epitomized by the musical magic of Kircher, Fludd, and even the Masonic Mozart, is still in evidence in the avant-garde music of the twentieth century. The numerological subtext of Schoenberg's music is an obvious connection to that tradition, and so is Hindemith's cosmological symbolism in *Die Harmonie der Welt*.

Another contemporary composer who can trace his conceptual lineage to that source is Karlheinz Stockhausen, whom many (not least the composer himself) have at one time or another perceived to be Schoenberg's principal heir, or the next musical giant after him. Born in 1928, Stockhausen, like so many composers of his generation, began his musical life as a disciple of the twelve-tone school, and he soon expanded into computer-generated music. His mature work has a deeply spiritual intent, although it is far from being religious in any conventional sense. Stockhausen closely resembles Schoenberg in at least one way, and that is in his unapologetic belief in his own genius. Although he has not formally set himself at the center of a worshipful cult in the way that Schoenberg did, he nonetheless clearly conceives of himself as a prophet of the direction that music should take.

Stockhausen's masterwork, a cycle of seven operas (one for each day of the week), called *Licht*, commences with *Donnerstag aus Licht* (Thursday from Light), about a trumpet-playing astronaut-angel named Michael, who flies through the heavenly spheres and returns to engage in *mano a mano* combat with Lucifer. In the finale, Michael turns to the audience and sings:

I became a HUMAN
to see myself and GOD the Father
as a human VISION,
to bring celestial music to humans
and human music to the celestial beings,
so that Man may listen to GOD
and GOD may hear his children.

As Michael's song suggests, Stockhausen closely follows Boethius's division of music into the congruent realms of *musica mundana*, *musica humana*, and *musica instrumentalis*, even to the point of using the same terms. In a talk for his disciples given at his home in Cologne, the composer articulated his conception of the sympathetic vibrations between instrumental music and human music: "Each of us is, as you know, a person with many levels. . . . I have a sexual center, three vital centers, two mental centers, and a suprapersonal center. . . . I can set my sexual center in vibration with a certain sort of music, but with another music I can set my supranatural center in vibration. . . . Hence it is naturally better if one hears music that draws one up higher than one is by nature."

He goes on to discuss the healing powers of music in terms almost identical with the descriptions in Pythagorean legend: "If for instance one is sick—say, we are too nervous or fearful or aggressive or tired of life—one can cure such sicknesses with music. . . . Very few realize that every one of us basically needs music for self-healing. Normally people drink coffee to become lively again. Few are wise enough and know the exact music by which one is always inwardly refreshed and dancing; a few just know that with certain pieces of Stockhausen, their ideas come ten times faster."

Then in classic Boethian fashion Stockhausen brings *musica mundana* into the equation: not only does instrumental music vibrate sympathetically with human music, it also provides the transcendental link between man and the cosmos:

"Then there naturally comes the next step, which religion also originally strove for, namely to bring ourselves through music into relationship with that which we cannot grasp with the understanding, but which we can feel; with the supernatural, with that which gives life to the whole universe—with God, the Spirit who holds everything together, all the galaxies, all the solar systems and planets, and also every single one of us on this little planet."

The concepts embodied by those words, aside from the astronomical particulars, come straight out of the Golden Verses of Pythagoras. Thus, as T. S. Eliot says, the end of all our exploring is to arrive where we started, an appropriately metaphysical passage for a metaphysical concept. It has been a progression from belief to theme: what was for the Pythagoreans inexpressible yet implicitly believed—indeed more than believed, what constituted the very essence of being human and a sensible part of the phenomenal universe—now exists only in formal expressions, and is believed only in the most self-conscious and artificial sense.

Stockhausen attempts to be the model Pythagorean, but one never believes for an instant that it is anything more than an attitude: not necessarily an insincere one, yet very far from being a spontaneous and profoundly held conviction. When he enumerates the centers of his self, one somehow knows that they will come to seven, the number of the celestial spheres and of the notes in a major scale. True, that requires three vital centers (whatever they may be) and two mental centers, where one each might have been thought to suffice, but it adds up to a "good" number. If seven had been too few, we may be sure that the total would somehow have been made to come to twelve, the number of the chromatic scale, a number with a pedigree. Yet for all his insistence on the Spirit and spirituality, Stockhausen still places his emphasis on the self, and that is what makes him an utterly modern man and alienated from the Pythagorean ideal. For in the great tradition the key is to find one's center outside of one's self, in the whole cosmos, paradoxically to become center-less; and that, it would seem, is the very last thing that mankind now is capable of.

In pursuing the concept of the musical universe from the first notes of Western music to the latest electronic screech, we have traced its gradual passage from vitality to sterility, from substance to form. Like some other great themes of Western civilization—the idea of God, or the idea of justice—it has not thrived in a world that is ever more insistent on the here and now, which believes only in what it knows. Yet now that science permits us actually to hear the soundtrack of the cosmos, in the form of random blips and howls picked up by radio telescopes, how we long for silence.

Bibliography

Abraham, Gerald. *The Concise Oxford History of Music*. London, 1979.

Adorno, Theodor. *Prisms*. Translated by Samuel and Shierry Weber. Cambridge, 1981.

———. *In Search of Wagner*. Translated by Rodney Livingstone. Trowbridge, 1981.

Allen, Reginald E. *Greek Philosophy: Thales to Aristotle*. New York, 1966.

Ammann, Peter J. "The Musical Theory and Philosophy of Robert Fludd." *Journal of the Warburg and Courtauld Institutes* 30 (1967), pages 198–227.

Armitage, Merle, ed. *Schoenberg*. New York, 1937.

Babbitt, Milton. *Words about Music*. Edited by Stephen Dembski and Joseph N. Straus. Madison, 1987.

Bacon, Francis. *New Atlantis*. Oxford, 1974.

Barker, Andrew, ed. *Greek Musical Writings, Volume I: The Musician and His Art*. Cambridge, 1984.

———. *Greek Musical Writings, Volume II: Harmonic and Acoustic Theory*. Cambridge, 1989.

Berger, Karol. *Theories of Chromatic and Enharmonic Music in Late 16th Century Italy*. Ann Arbor, 1976.

Bianconi, Lorenzo, *Music in the Seventeenth Century*. Cambridge, 1987.

Blom, Eric, ed. *The New Everyman Dictionary of Music*. New York, 1988.

Blume, Heinrich. *Two Centuries of Bach*. London, 1950.

Boethius. *The Principles of Music*. Translated by Calvin Martin Bower. Ann Arbor, 1967.

Chailley, Jacques. *The Magic Flute, Masonic Opera*. New York, 1971.

Cicero. *Tusculanaraum Disputationum*. Edited by Frank Ernest Rockwood. Boston, 1903.

Cornford, F. M. *Plato's Cosmology*. London, 1937.

———. *Plato and Parmenides*. London, 1939.

Courcy, G. I. C. de. *Paganini the Genoese*. Norman, Okla., 1957.

Curl, James Stevens. *The Art and Architecture of Freemasonry*. London, 1991.

Dahlhaus, Carl. *Schoenberg and the New Music*. Translated by Derrick Puffett and Alfred Clayton. Cambridge, 1987.

D'Angelo, James P. "Tonality and Its Symbolic Associations in Paul Hindemith's Opera *Die Harmonie der Welt*." Ph.D. diss., New York University, 1983.

Dobbs, Betty Jo Teeter. *The Foundations of Newton's Alchemy*. Cambridge, 1975.

Dyer, Denys. *The Stories of Kleist: A Critical Study*. New York, 1977.

Einstein, Alfred. *Music in the Romantic Era*. New York, 1947.

Eisenbichler, Konrad, and Pugliese, Olga Zorzi, eds. *Ficino and Renaissance Neoplatonism*. Ottawa, 1986.

Fauvel, John; Flood, Raymond; Shortland, Michael; and Wilson, Robin, eds. *Let Newton Be!* Oxford, 1988.

Flaubert, Gustave. *Madame Bovary*. Translated by Marx Aveling. New York, 1950.

Franklin, Don O., ed. *Bach Studies*. Cambridge, 1989.

Freden, Gustaf. *Orpheus and the Goddess of Nature*. Göteborg, 1958.

Gafurius, Franchinus. *Practica Musica*. Edited by Irwin Young. Madison, 1969.

Godwin, Joscelyn, *Harmonies of Heaven and Earth*. Rochester, Vt., 1987.

———. *Athanasius Kircher*. London, 1979.

———. *Robert Fludd*. London, 1979.

———, ed. *Music, Mysticism, and Magic*. London, 1986.

Guthrie, Kenneth Sylvan, trans. and comp. *The Pythagorean Sourcebook and Library*. Grand Rapids, 1987.

Guthrie, W. K. C. *Orpheus and Greek Religion*. New York, 1966.

Haskell, Harry. *The Early Music Revival: A History*. New York, 1988.

Hindemith, Paul. *The Craft of Musical Composition*. Translated by Arthur Mendel. New York, 1945.

———. *Introductory Remarks for the New Version of the Song Cycle* Das Marienleben. Translated by Arthur Mendel. New York, 1948.

Hogwood, Christopher. *Handel*. London, 1984.

Jones, George Thaddeus. *Music Theory*. New York, 1974.

Keyte, Hugh. Notes for *Una "Stravaganza" dei Medici*. Andrew Parrott and the Taverner Consort, Taverner Choir, and Taverner Players. EMI CDC 7 47998 2.

Koestler, Arthur. *The Sleepwalkers*. New York, 1959.

Kuhn, Thomas S. *The Structure of Scientific Revolutions*. Chicago, 1962.

La Croix, Richard R., ed. *Augustine on Music*. Lewiston, N.Y., 1988.

Lane Fox, Robin. *Pagans and Christians*. New York, 1986.

Lerner, Ralph, ed. and trans. *Averroës on Plato's Republic*. Ithaca, 1974.

Lipman, Edward A. *Musical Thought in Ancient Greece*. New York, 1964.

Macrobius. *Commentary on the Dream of Scipio*. Translated by William Harris Stahl. New York, 1952.

Maniates, Maria Rika. *Mannerism in Italian Music and Culture, 1530–1630*. Chapel Hill, 1979.

McGuire, J. E., and Rattansi, P. M. "Newton and the 'Pipes of Pan.'" *Notes and Records of the Royal Society of London* 21, pages 108–43.

Mei, Girolamo. *Letters on Ancient and Modern Music to Vincenzo Galilei and Giovanni Bardi*. Edited by Claude V. Palisca. American Institute of Musicology.

Mellers, Wilfrid. *Bach and the Dance of God*. New York, 1981.

Metastasio, Pietro. *Opere*. Florence, 1819.

Meyer-Baer, Kathi. *Music of the Spheres and the Dance of Death*. Princeton, 1970.

Palisca, Claude V. *Humanism in Italian Renaissance Musical Thought*. New Haven, 1985.

Peyser, Joan. *The New Music*. New York, 1971.

Pirrotta, Nino. *Music and Culture in Italy from the Middle Ages to the Baroque*. Cambridge, 1984.

Plato. *The Republic*. Edited by F. M. Cornford. Oxford, 1941.

————. *Timaeus* and *Critias*. Edited by Desmond Lee. London, 1971.

————. *Works*. Translated by Thomas Taylor. London, 1804.

Pulver, Jeffrey. *Paganini: The Romantic Virtuoso*. New York, 1970.

Reese, Gustave. *Music in the Middle Ages*. New York, 1940.

————. *Music in the Renaissance*. New York, 1954.

Reich, Willi. *Schoenberg: A Critical Biography*. Translated by Leo Black. London, 1971.

Rosenau, Helen, ed. *Boullée's Treatise on Architecture*. London, 1953.

Stahl, William Harris; Johnson, Richard; and Burge, E.L. *Martianus Capella and the Seven Liberal Arts*. New York, 1977.

Strunk, Oliver, ed. and comp. *Source Readings in Music History: From Classical Antiquity Through the Romantic Era*. New York, 1950.

Tatlow, Ruth. *Bach and the Riddle of the Number Alphabet*. Cambridge, 1991.

Testi, Flavio. *La musica italiana nel Seicento*. Milan, 1970.

Tillyard, E. M. W. *The Elizabethan World Picture*. New York, 1967.

Vickers, Brian, ed. *Occult and Scientific Mentalities in the Renaissance*. Cambridge, 1984.

Walker, D. P., *Spiritual and Demonic Magic*. London, 1958.

––––––. *Studies in Musical Science in the Late Renaissance*. London, 1978.

Weinstock, Herbert. *Donizetti*. New York, 1979.

Xenakis, Iannis. *Formalized Music*. Bloomington, 1971.

Yates, Frances A. *The French Academies of the Sixteenth Century*. London, 1947.

––––––. *Giordano Bruno and the Hermetic Tradition*. Chicago, 1964.

Index

A

abortion, 7
Abraham, 134
a cappella singing, 90
Adams, John, 230–231
Aeneid (Virgil), 20
aestheticism, 4, 58, 184
afterlife, 55, 63, 65, 67–68
Age of Reason, 4
Age of Reason, The (Paine), 184
Aglaophemus, 118
Agricola de Metallis, 160
Aida (Verdi), 208
alchemy, 86, 140, 141, 155, 160,
     167, 214, 217, 219
Alexander VI, Pope, 124–125
Alexander Polyhistor, 39
Allgemeine musikalische Zeitung, 190–
     191
Almagest (Ptolemy), 61
Anaxilas, 68
Anaximander, 37–38
Anaximenes, 38
angels, 71, 111–112
*Annotations Physical, Mathematical,
     and Theological* (Gregory), 163
anti-Semitism, 208

Ape of Nature, The (Fludd), 129–131
Aquinas, Saint Thomas, 77
"Arcades" (Milton), 112–113
Archimides, 28
Archippus, 24, 27
Archytas, 72
Arezzo cathedral, 81
arias, 15, 102, 182–183
Aristotle, 28, 30, 38–39, 40, 51, 60–
     61, 71, 162
Aristoxenus, 78, 90
arithmetic, 72, 74, 88
armillary sphere, 48–49, 50
"armonia delle sfere, L'," 103–105,
     108
Arnold, Matthew, 110
"Ars gratia artis," 58, 225
ars nova, 84–85
Ars novae musicae (Jean de Muris),
     83–84
*Art and Architecture of Freemasonry,
     The* (Curl), 168–169
"Art for art's sake," 58, 225
Art of the Fugue, The (Bach), 183, 192
arts:
     decorative, 110
     nature and, 90
     plastic, 194, 233

THE FUTURE OF LIFE

Edward O. Wilson

Winner of the BP Natural World Book Prize 2002
'The world's greatest living writer on science' *The Times*

From one of the world's most influential scientists, an impassioned
call for quick and decisive action to save the Earth's
rich biological heritage.

'The greatest natural history expert of our age . . . this is a
brilliantly constructed analysis of our planetary woes, a much
needed response to those who think our environments are safe and
secure from global degradation' *Observer*

'Dawkins and his many imitators may enthral today's crowds;
Wilson will enthral tomorrow's . . . Wilson is among the greatest
living writers of English prose . . . [his] greatness also lies in the
quality of his thought . . . a tough and gripping polemic'
Literary Review

'[Wilson] is certainly the most eloquent, measured and expert
writer we have on biological science . . . in The Future of Life,
Wilson does not disappoint: it by far the best argued, and certainly
the best written book I have read on the topic' *Daily Mail*

'A grippingly detailed account . . . as always with Wilson, the book
is a great pleasure. The wonderful range of detailed and often
arcane knowledge on the natural world is matched by the limpid
clarity of his prose' *Independent on Sunday*

'There's a new Darwin. His name is Edward O. Wilson'
Tom Wolfe

Abacus
978-0-349-11579-5

FASTER

The Acceleration of Just About Everything

James Gleick

Do you hit the 'door close' button because the lift doors are taking too long to shut? Did you know we work longer, commute longer, shop longer and sleep 20% less than we did a century ago? We are obsessed with making more time, yet a microwave oven saves just four minutes a day — the average amount of time we spend having sex or fillling in government forms.

With acute insight and mordant will, James Gleick dissects our unceasing struggle to squeeze as much as we can into the 1,440 minutes of each day. From one-minute bedtime stories to Federal Express, from multi-tasking to the double-edged benefits of e-mail, *Faster* is the perfect self-help book for the twenty-first century, a description of our lives and an answer to our fundamental complaint that there is never enough time.

'*Faster* will make you think . . . if you want to understand why your To Do list might be brimming over, you'll add *Faster* to it'
Financial Times

'A fascinating meditation about the modern definition of time that helps to define the age we live in'
Daily Telegraph

'There is no way a reviewer can do justice to the density of anecdote and fact here . . . a delightful read'
Times Higher Education Supplement

James Gleick, a science writer with a healthy pop sensibility, has written a highly readable dissection of our speed-obsessed age'
The Face

Abacus
978-0-349-11292-3

THE QUARK AND THE JAGUAR

Adventures in the Simple and the Complex

Murray Gell-Mann
Winner of the Nobel Prize for Physics

'Teeming with insights and intelligent comment'
Independent

From one of the century's greatest scientists comes this unique, highly personal vision of the connections between the basic laws of physics and the complexity and diversity of the natural world. *The Quark and the jaguar* is an irresistibly engaging and rewarding introduction to the life's work of physicist, polymath and Nobel Laureate Murray Gell-Mann.

'Reports from the cutting edge, presented with eloquence and style . . . (it gives) the dizzying sense that science is poised on the brink of a new world of discovery; a world stranger and still more beautiful than anything imagined yet'
A. C. Grayling, *Financial Times*

'Every once in a while a physicist, with a gift for writing and an empathy for the non-scientific mind, reports back like an anthropologist from Mars. Gell-Mann has written just such a book'
John Cornwell, *Spectator*

'Stimulating and provocative'
Carl Sagan

'Fascinating'
Graham Ross, THES

Abacus
978-0-349-10649-6

THE TIPPING POINT

How little things can make a big difference

Malcolm Gladwell

'In the words of the now-famous book that everybody is reading, it
reaches a kind of tipping point and people kind of get it' Bill
Clinton, White House press conference, 28 June 2000

In this brilliant and original book, Malcolm Gladwell explains and
analyses the 'tipping point', that magic moment when ideas, trends
and social behaviours cross a threshold, tip and spread like wildfire.
Taking a look behind the surface of many familiar occurrences in
our everyday world, Gladwell explains the fascinating social
dynamics that cause rapid change.

'Provides some profoundly suggestive arguments and insights . . .
Fascinating' *Sunday Telegraph*

A fascinating book that makes you see the world in a different way'
Fortune

A wonderfully offbeat study of that little-understood phenomenon,
the social epidemic'
Daily Telegraph

'Genuinely fascinating and frequently startling . . . The kind of
book from which you'll be regaling your friends with intriguing
snippets for weeks to come'
Scotland on Sunday

Abacus
978-0-349-11346-3

To buy any of our books and to find out
more about Abacus and Little, Brown, our authors
and titles, as well as events and book clubs,
visit our website

www.littlebrown.co.uk

and follow us on Twitter

@AbacusBooks
@LittleBrownUK

To order any Abacus titles p & p free in the UK,
please contact our mail order supplier on:

+ 44 (0)1832 737525

Customers not based in the UK should contact
the same number for appropriate postage
and packing costs.